Sustainable Practice for the Facilities Manager

D1440505

....ı 01233 6549

To Helen

Sustainable Practice for the Facilities Manager

Sunil Shah

Blackwell Publishing

© 2007 Sunil Shah

Blackwell Publishing Ltd
Editorial Offices:
Blackwell Publishing Ltd, 9600 Garsington Road, Oxford OX4 2DQ, UK
  Tel: +44 (0)1865 776868
Blackwell Publishing Inc., 350 Main Street, Malden, MA 02148-5020, USA
  Tel: +1 781 388 8250
Blackwell Publishing Asia Pty Ltd, 550 Swanston Street, Carlton, Victoria 3053, Australia
  Tel: +61 (0)3 8359 1011

The right of the Author to be identified as the Author of this Work has been asserted in accordance with the Copyright, Designs and Patents Act 1988.

All rights reserved. No part of this publication may be reproduced, stored in a retrieval system, or transmitted, in any form or by any means, electronic, mechanical, photocopying, recording or otherwise, except as permitted by the UK Copyright, Designs and Patents Act 1988, without the prior permission of the publisher.

First published 2007 by Blackwell Publishing Ltd

ISBN-13: 978-1-4051-3557-3
ISBN-10: 1-4051-3557-3

Library of Congress Cataloging-in-Publication Data
Shah, Sunil
  Sustainable practice for the facilities manager/Sunil Shah.
    p. cm.
  Includes bibliographical references and index.
  ISBN-13: 978-1-4051-3557-3 (pbk.: alk. paper)
  ISBN-10: 1-4051-3557-3 (pbk.: alk. paper)
1. Facility management. 2. Sustainable development. I. Title.

  TS155.S3966 2006
  658.2–dc22

2006012518

A catalogue record for this title is available from the British Library

Set in 10/13 pt Palatino
by Newgen Imaging Systems (P) Ltd., Chennai, India
Printed and bound in India
by Replika Press Pvt, Ltd

The publisher's policy is to use permanent paper from mills that operate a sustainable forestry policy, and which has been manufactured from pulp processed using acid-free and elementary chlorine-free practices. Furthermore, the publisher ensures that the text paper and cover board used have met acceptable environmental accreditation standards.

For further information on Blackwell Publishing, visit our website:
www.blackwellpublishing.com

Contents

List of Case Studies

A series of case studies have been used throughout this book to provide relevant examples of both good and poor practices. A list of those organisations that have provided information in these examples are described below:

Chapter 2

2.1 – Green service management

Oxfam GB – Ethical purchasing

Authorised by Rachel Wilshaw, Purchasing Strategy Manager

Oxfam GB is a development, relief, and campaigning organisation that works with others to find lasting solutions to poverty and suffering around the world.

Oxfam's Beliefs

❑ The lives of all human beings are of equal value;
❑ In a world rich in resources, poverty is an injustice which must be overcome;
❑ Poverty makes people more vulnerable to conflict and natural calamity; much of this suffering can be prevented, and must be relieved;
❑ People's vulnerability to poverty and suffering is increased by unequal power relations based on, for example, gender, race, class, caste and disability; women, who make up a majority of the world's poor, are especially disadvantaged;
❑ Working together we can build a just and safer world, in which people take control over their own lives and enjoy their basic rights;
❑ To overcome poverty and suffering involves changing unjust policies and practices, nationally and internationally, as well as working closely with people in poverty.

2.4 – Corporate responsibility management systems

BAE Systems, Basildon – Innovative approach secures ISO14001 certification

Authorised by David John, Operations Director

Since certification was achieved in 2000, the site has become part of the SELEX Sensors and Airborne Systems Company – a joint venture with Finmeccanica. The company specialise in world-class integrated sensor solutions, surveillance, protection, tracking, targeting and imaging systems. The Basildon site complex covers two main sites housing over 1500 staff.

2.5 – Corporate responsibility reporting

Gap – Sustainability report

Taken from www.gap.com website

Gap Inc. is one of the world's largest specialty retailers, with more than 3000 stores and fiscal 2005 revenues of $16 billion. We operate four of the most recognised apparel brands in the world – Gap, Banana Republic, Old Navy and Forth & Towne. A public company, Gap Inc. is traded on the New York Stock Exchange and is listed on the Calvert Social Index, Domini 400 Social Index, Dow Jones Sustainability Index, and the FTSE4Good US 100 and FTSE4Good Global Indices.

Chapter 3

3.1 – Life cycle facilities approach

Carillion – Corporate sustainability strategy

Authorised by Quentin Leiper, Aneysha Minocha and Stuart Mee, Sustainability Team

Carillion provides a broad range of business, transport and construction services to commercial and public sector clients in the United Kingdom, Sweden, Canada and the Middle East, with a turnover of £2billion and around 17 000 employees (13 500 UK, 3500 overseas). Carillion's mission is to make tomorrow a better place. They will do this by being the leader in integrated solutions for infrastructure, building and services, and by acting according to our values of openness, innovation, mutual dependency, collaboration, sustainable profitable growth and professional delivery.

3.3 – Design

National Trust – Heelis central office, Swindon

Authorised by John Seaber and Liz Adams, Facilities Team

The National Trust works to preserve and protect the coastline, countryside and buildings of England, Wales and Northern Ireland. The organisation is a charity and is completely independent of the UK Government, relying for income on membership fees, donations and legacies, and revenue raised from commercial operations. They have 3.4 million members and 43 000 volunteers. More than 12 million people visit pay for entry properties, while an estimated 50 million visit open air properties. The National Trust protect and open to the public over 300 historic houses and gardens and 49 industrial monuments and mills, and also look after forests, woods, fens, beaches, farmland, downs, moorland, islands, archaeological remains, castles, nature reserves, villages.

3.5 – Refurbishment, fit-out and project management

Building Design Partnership – London studio

Authorised by Trevor Butler, Director of Sustainability

BDP is a multi-discipline practice of architects, designers and engineers, aiming to provide comprehensive, integrated design services for the built environment. The mission is to be collaborative, working in partnership with customers, colleagues in construction and the communities we impact. The company's aim is to be ethical and responsible to all stakeholders, believing that good design can create substantial economic and social value.

BDP designs about 2% of all the new, non residential building (and some of the housing) in the UK each year, around £450 million in construction costs, with an additional 10% of work overseas. There are over 800 staff in the United Kingdom in 10 locations with 140 more in Ireland and in France in BDP International. BDP has an annual fee income of over £60 million and are consistent major design award winners, with over 300 to date.

Refurbishing housing – An energy efficient house

Authorised by Dave Hampton

Chapter 4

4.1 – Maintenance

EMS – A series of case studies

Authorised by Paul Foxcroft, Managing Director

Formally known as Environmental Monitoring Services, ems, was established as a consultancy in 1986 and now supports clients to achieve practical solutions to help them demonstrate compliance with legislation and raise their profile within their organisation. The company has made the decision not to provide any remedial services or to endorse the products or services of other organisations. Their independence ensures clients receive balanced advice that can help make better informed decisions about the working environment and service providers.

4.2 – Operation

Carillion – Swindon and Marlborough NHS Trust

Authorised by Quentin Leiper, Aneysha Minocha and Stuart Mee, Sustainability Team

About the Author

Sunil Shah   Sunil has over ten years' experience within the built environment, reviewing life cycle environmental and social impacts from facilities and where benefits can be made and impacts minimised. Sunil has experience in a number of sectors, providing strategic consultancy support for clients including GlaxoSmithKline, BP, BAA, HM Prison Service, Pfizer and London Fire Brigade.

A career which started as a geologist led onto an environmental scientist position with a manufacturing organisation helping achieve ISO14001 and provide the monitoring for many of the stack and emissions from site. Sunil joined Johnson Controls Consultancy Group in the late 1990s where he helped develop and group the sustainability team prior to leaving in 2004 to join Jacobs.

Sunil has written a number of articles and regularly provides training to facilities managers on the current pitfalls and easy to implement means to deliver sustainable practices.

Preface

Buildings impact upon every fact of our lives – from work, play and housing, they provide the basic infrastructure and few are aware of how they are developed and operated.

Buildings are playing a greater role in our lives, affecting society and the environment. Issues such as work–life balance, climate change, water shortages, fair pay and human rights have all been front page news over the past few years. Awareness of the issues has been raised alongside the impacts of buildings and the desire from individuals and business to look for alternatives.

Buildings are replenished at a slow rate, with many of those standing and in use well over 100 years old. The current churn rates – the rate at which buildings are replaced – are less than five per cent in general, requiring a greater focus to be placed upon existing buildings.

This book focuses upon the role the facilities manager has during the life cycle of the building, from the initial briefing stages, though design, construction, operation, and refurbishments. The aim is to close the knowledge loop between the design and operation of the building – both to provide feedback to the designers on good and poor practices, as well as including the facilities manager or consultant to be involved at the initial stages to challenge the design.

Throughout these stages, the ability to incorporate sustainability and the understanding of the building to meet end user requirements should be promoted. Although many modern buildings do include elements of sustainability, this book looks at how this can be maximised in a consistent process to deliver a better value building for the same price.

My goal throughout this book has been to provide practical guidance and information which can be implemented to integrate sustainability into the day to day activities of facilities managers.

Abbreviations

ACCA – Association of Certified Chartered Accountants
ASHRAE – American Society of Heating, Refrigeration and Air Conditioning
Engineers

B(E)MS – Building (Energy) Management System
B2B – Business to Business
BAP – Biodiversity Action Plan
BCM – Bromochloromethane
BDP – Building Design Partnership
BIFM – British Institute of Facilities Management
BITC – Business in the Community
BRE – Building Research Establishment
BREEAM – Building Research Establishment Environmental Assessment
Methodology
BS – British Standard

CAA – Clean Air Act
CAFM – Computer Aided Facilities Management
CEO – Chief Executive Officer
CERCLA – The Comprehensive Environmental Response, Compensation, and
Liability Act Superfund
CESMP – Construction Environmental and Social Management Plan
CFC – Chlorofluorocarbon
CFM – Chartered Facilities Manager
CFR – Code of Federal Regulations
CHP – Combined Heat and Power
CIBSE – Chartered Institution of Building Services Engineers
CIRIA – Construction Industry Research Association
CIS – Co-operative Insurance Society
CO_2 – Carbon Dioxide
COSHH – Control of Substances Hazardous to Health
CPD – Continual Professional Development
CRMS – Corporate Responsibility Management System
CWA – Clean Water Act

DBFO – Design-Build-Finance-Office
DEH – Department of the Environment and Heritage
DIY – Do It Yourself

DJSI – Dow Jones Sustainability Index
EBIT – Earnings before Interest and Tax
EC – European Community
ECJ – European Court of Justice
EHS – Environment, Health and Safety
EIA – Environmental Impact Assessment
EMAS – Eco-Management and Audit Scheme
EMS – Environmental Management System
EPA – US Environmental Protection Agency
EPBD – Energy Performance of Buildings Directive
EPCA – Environmental Pollution Control Act
EPD – Environmental Protection Department
ES – Environmental Statement
EU – European Union

FM – Facilities Management
FMA – Facilities Management Australia
FSC – Forest Stewardship Certification
FTE – Full Time Equivalent

GABS – Global Alliance for Building Sustainably
GBTool – Green Building Tool
GDP – Gross Domestic Product
GFT – Global Fortune
GHG – Green House Gas
GRI – Global Reporting Initiative

HCFC – Hydrochlorofluorocarbon
HK-BEAM – Hong Kong Building Environmental Assessment Method
HMIP – Her Majesty's Inspectorate of Pollution

IAQ – Indoor Air Quality
IFM – Integrated Facilities Management
IFMA – International Facilities Management Association
IiP – Investors in People
ILO – International Labour Organisation
IMS – Integrated Management System
IPCC – Integrated Pollution Prevention and Control
IPCC – Intergovernmental Panel on Climate Change
ISO – International Organisation for Standardisation
IT – Information Technology

KPI – Key Performance Indicator
kW – Kilowatt
kWh – Kilowatt Hours

LCA – Life cycle analysis
LEED – Leadership in Energy and Environmental Design

LPG – Liquefied Petroleum Gas

M&E – Mechanical and Electrical
M&T – Monitoring and Targeting
MD – Managing Director
MEPI – Measuring Environmental Performance of Industry
MFD – Multi-functional Devices
MRF – Materials Recycling Facility
MSC – Master of Science

NABERS – National Australian Building Environmental Rating System
NGO – Non Governmental Organisation
NHS – UK National Health Service
NIA – Net Internal Area
NPI – Normalised Performance Index
NRA – National Rivers Authority
NT – National Trust

O&M – Operating and Maintenance
ODS – Ozone Depleting Substance
OGC – UK Office of Government Commerce
OJEU – Official Journal of the European Union
OPA – The Oil Pollution Act

P(P)M – Planned (Preventative) Maintenance
p.a. – per annum
PBB – Polybrominated Biphenyls
PBDE – Polybromided Diphenyl Ethers
PEFC – Programme for the Endorsement of Forest Certification
PFI – Private Finance Initiative
POE – Pre / Post Occupancy Evaluation
PR – Public Relations
PV – Photovoltaic

RCRA – The Resource Conservation and Recovery Act
RFI – Request for Information
RFP – Request for Proposal
RICS – Royal Institution of Chartered Surveyors
ROHS – Restriction of use of Certain Hazardous Substances

SA – Social Accountability
SAP – Sustainability Action Plan
SBS – Sick Building Syndrome
SCP – Sustainable Construction Potential
SDA – Sewerage and Drainage Act
SEPA – Scottish Environment Protection Agency
SIGMA – Sustainability - Integrated Guidelines for Management
SME – Small and Medium Sized Enterprise

SMR – Sustainability Management Representative
SMS – Sustainability Management System
SOG – Sustainability Operations Group
SRI – Socially Responsible Investment
SWDA – The Solid Waste Disposal Act
SWMP – Site Waste Management Plan

TBL – Triple Bottom Line
TSCA – The Toxic Substances Control Act

UK – United Kingdom
UN – United Nations
UNEP – United Nations Environment Programme
UNFCCC – United Nations Framework Convention on Climate Change
UPS – Uninterrupted Power Supply
US – United States of America

VFM – Value for Money
VOC – Volatile Organic Compounds

WC – Water Closet
WEEE – Waste Electrical and Electronic Equipment Directive
WLV – Whole Life Value
WRI – World Resources Institute
WSSD – World Summit on Sustainable Development
WWF – World Wildlife Fund

Introduction

Both sustainability and facilities management (FM) are substantial topics in their own right and have been covered in detail through a number of books and research projects over the past few years. This has been in recognition of the major impacts both subjects have on the way we lead our lives and how we work. Having developed dramatically over the past decades, the concepts of the two subjects are still maturing and, as such, are often blurred and can cause confusion for many not involved in the intricacies.

One downside of the confusion is to make the subject appear more complex than it really is. Another is to put off many of those who can contribute and provide real benefits.

As such, this book does not try to cover the two subject areas in any great detail. Instead it captures the salient points relevant to the incorporation of sustainability criteria within the lifecycle of buildings, with an onus placed upon delivering an improved operationally performing facility. Coupled with this are a series of additional references and websites provided for further reading, which I have found particularly useful. Again this is the tip of the iceberg and certainly does not represent all the sites available.

The integration of the two subjects into sustainable facilities management is a relatively new area. The role of FM has been taken up throughout the design, construction and operation of the built environment, with a function to provide the feedback of knowledge on effective management of the facility. The premise is very much that the focus of facilities should be driven by the end-user, and the operational performance, and therefore the facilities manager should be represented at the table during such discussions at any point during the lifecycle where significant changes occur.

In order to achieve this, it is not as simple as being available and present at the meetings. The facilities manager has to speak the same language, discuss and engage with the business and the project team to influence the outcomes. Knowledge and skills of the building lifecycle are as important as understanding the end-user needs. Unfortunately, this is a hurdle for many facilities managers and are commonly not invited due to lack of knowledge and skills. I have personally witnessed such discussions and actions many times.

It is this challenge that has inspired this book. The scope will review the whole building lifecycle, from initial briefing through to final disposal and the role FM has to play in this. Due to the size of this area, there will be some limitations in the sectors chosen, focusing on the commercial/office buildings, housing and retail.

There is also a focus on legislation covering the United Kingdom, European Union, Australia, Hong Kong, United States, Canada and Singapore.

Throughout the book, a set of 14 sustainability categories are referred to and encompassed within a set of processes and methodologies including templates to provide means to readily understand the impacts and opportunities and convey them to the business and project team. Whilst these processes and methodologies have been described for a few sectors, they can be used across all sectors.

The contents will be split into four sections to make use and navigation easier, and also devote distinct areas of the book to review a myriad of topics the subject covers. Each section commences with a summary to highlight the key points covered and the lessons learnt, enabling the salient points to be digested quickly. The sections are described below, with their respective chapters:

❑ Sustainable Development and Facilities Management – The first chapter acts to introduce the reader to the concepts of Facilities Management and Sustainability, and how the components fit together;
❑ Sustainable Business Management – Set the introduction in context with the business operations, drivers and wider issues that the Facilities Manager will need to accommodate. This will include the wider business functions including management systems, procurement processes and the business planning cycle;
❑ Facilities Life cycle – The incorporation of sustainability criteria into capital project plans, based upon the front end stages of the building life cycle – from briefing through to handover. This will enable easier understanding from the FM knowledge base. Practical means of change with case studies and additional reading has been provided.
❑ Operation of the Facility – The incorporation of sustainability criteria into day to day FM, based upon the operation, maintenance and occupation of the facility. Good practice measures and a number of templates have been provided.

1 Sustainable development and facilities management

This opening chapter will introduce the concepts of facilities management (FM) and sustainable development both as individual subjects in their own right and what it means to deliver sustainable facilities management.

Both topics are significant pieces individually, with a range of books, publications and websites devoted to each subject. The sections covered in this chapter do not try to replace the information contained in these publications but merely represent the main points which will affect the aspects related to sustainable facilities management and are of relevance in later chapters of the book. Further information on both of these topics can be found in the Map of Organisations and list of websites at the end of this book.

The final section pulls together the two aspects and represents a series of 14 topic headings to define sustainability within the FM and property activities, which are used throughout this book. The use of these 14 categories will help to simplify the language used and help to identify where benefits can be provided and implemented.

1.1 The growing age of facilities management

Facilities management (FM) means many things to many people – at one end it leaves a core team to manage the business with a series of contractors to perform the non-core activities. At the other, it is the in-house provision of most activities with limited out-tasking such as cleaning.

The definitions of FM are still under debate, although recently the global community is coming to a consensus on the main aspects. The role of FM in business is undergoing a period of global growth with recognition of the benefits effective delivery can achieve. The industry is still maturing with a growing list of services defined as non-core and within the remit of FM to provide.

Section learning guide

This section provides an overview of FM, its growth globally and the challenge in trying to define its role in the changing business requirements. Included are the various types of FM implemented:

❏ The development and growth of FM globally with the cultural fit to each society;
❏ The range of activities involved in delivering FM;
❏ Relationship between FM and business, and the inherent challenges;
❏ Differences across Europe, the United States and Asia based upon labour requirements and maturity of the FM model; and
❏ The future of FM and where the industry is going.

Key messages include:

❏ FM can provide real value as an integrated part of business activities; and
❏ The role of society and community is becoming an increasing focus for the FM industry.

1.1.1   Introduction

In the United States, FM started in the 1970s with the International Facility Management Association (IFMA) set up in 1980 with a number of organisations looking to train and manage staff involved at the interface between the workplace, staff and processes. The roll-out of the FM market globally has been slowly growing with the establishment of Euro FM in 1990 and the British Institute of Facilities Management (BIFM) in 1993.

There has been conjecture over the exact dates of when FM started, in part, due to the linkage with the term FM used by the IT and military sectors. More recently, the questions have concentrated on the scope and extent of what FM covers. It is recognised that FM includes a broad range of services, such as catering, cleaning, building management and maintenance, ground management and maintenance, security, postal, data management and IT, telecommunications, secretarial, health and safety, to organisations. It is also increasingly beginning to include human resources and finance functions as well.

1.1.2   Definitions of facilities management

FM can be provided by either an in-house team/department or by companies under contract.[1] There is an increasing tendency for organisations to contract out (outsource/out-task) non-core business activities to provide the best service at the lowest cost.

---

[1] Mintel Report, *Facilities Management – UK*, Mintel International Group Ltd, December 2005.

The breadth of FM is highlighted in the divergent views of some of the leading practitioners. None of these views are right or wrong – they merely highlight the changing needs of organisations.

The design, management and implementation of all non-core activities within an organisation, allowing the core activities to maximise cost efficiency.

John Davis, Managing Director, Facilities Recruitment Limited

A profession that encompasses multiple disciplines to ensure functionality of the built environment by integrating people, place, process and technology.

IFMA Board of Directors, 2004

A business practice that optimises people, process, assets and the work environment to support delivery of the organisation's business objectives.

FMA Australia's Glossary of FM Terms

What it means: Facilities management is the integration of multi-disciplinary activities within the built environment and the management of their impact upon people and the workplace.

The value it provides: Effective facilities management, combining resources and activities, is vital to the success of any organisation. At a corporate level, it contributes to the delivery of strategic and operational objectives demonstrating corporate social responsibility. On a day-to-day level, effective facilities management provides a safe and efficient working environment, which is essential to the performance of any business – whatever its size and scope.

The future of FM: Within this fast growing professional discipline, facilities managers have extensive responsibilities for providing, maintaining and developing myriad services. These range from property strategy, space management and environmental issues to building maintenance, administration and contract management. Facilities management will continue to expand and encompass further non-core activities including payroll, voice and data and IT, which have been traditionally managed outside of the FM discipline.

Ian R. Fielder, CEO, British Institute of Facilities Management

Facility management is a discipline that improves and supports the productivity of an organization by delivering all needed appropriate services, infrastructures, etc. that are needed to achieve business objectives.

CEN/TC 348 Facilities Management

The BIFM have defined five distinct areas of FM:[2]

(1) Large companies with in-house facilities teams that manage contracts with outsourced suppliers;

---

[2] British Institute of Facilities Management – http://www.bifm.org.uk

- Catering
- Cleaning
- Data management and IT
- Building management and maintenance
- Ground management and maintenance
- Security
- Procurement
- Project management
- Telecommunications
- Secretarial
- Postal
- Health and safety

Figure 1.1   FM service categories.

(2) External management suppliers which offer a range of outsourced services as a total one-stop shop;
(3) Smaller individual suppliers providing specific contracts for services such as cleaning or pest control;
(4) Product suppliers; and
(5) Consultants.

Although FM has been defined by the delivery of non-core services, numerous broad provision types have been developed by which firms can contract out some or all of their activities. Figure 1.1 describes some of the key FM service categories. Facilities management provision may either be organised across service areas or via a packaged portfolio of service areas. These incorporate the following:

Single service provision: Focus on the delivery of one particular type of service, such as cleaning, security, catering and maintenance.

Multiple or packaged service provision: Companies are now also providing an extensive range of services as a package. This means that a security firm may supply manned guards, burglar alarm systems, and/or electronic entry systems all as part of a single package, rather than, for example, manned guards alone. Once again, the focus is solely on delivery.

Management contracting: A mixture of delivery and management services are provided where client organisations will hire several management contractors, subsequently giving the client company overall management control. Meanwhile, management contractors will typically concentrate on the provision of a small number of service categories.

Integrated facilities management (IFM): At this juncture, client companies delegate total management of their facility requirements to service providers who manage and deliver an extensive range of services directly or by sub-contract(s). The increasing appeal of IFM companies to client organisations can be attributed to the fact that they focus their attention on the management of services, thus enabling the client organisation to avoid getting involved on a day-to-day basis.

Management agents: Here there is a focus on management alone where agents control provision by contracting the services of companies focusing on the delivery of facilities.

Infrastructure management: Involves the maintenance and management of bridges, railways, roads, and utilities (gas, water, and electricity).

Building operations and maintenance: Encompasses buildings and content management, maintenance plus plant and systems management, including electrical, fabric, mechanical apparatus, specialist equipment and grounds.

Business support services: Administration, finance, human resources management or procurement, and insurance are some of the categories of services in this area.

Support services: Encompasses a vast array of service streams such as bar management, catering, cleaning, courier services, furniture provision, interior environment, laundry, mail services, porterage, security, travel, library, and shops/retail, office and business support – essentially reprographics, printing, secretarial, reception and vending.

Property management: Essentially involves asset management, design/construction, disposals/acquisitions, space planning, project management, and relocation management.

1.1.3   Business challenges for FM

A significant driver for the FM market within private and public organisations is to reduce costs, promote efficiency and improve performance against an increasing complexity of organisational models and the application of new technologies. FM is acknowledged as contributing to the bottom line with an increased recognition of the importance of quality and robust processes, be it risk management or cost reduction. There is a growing recognition that FM is for the support of the primary purpose of an enterprise.

The perception of FM however is being widely considered as a commodity service for cost reduction as opposed to a strategic discipline capable of working with an organisation to provide real value.

A report published by Frost and Sullivan[3] captured the movement towards outsourcing for the North American market specifically, but is applicable globally. It concludes that outsourcing is the smart way to cut costs, increase productivity, and focus on revenue-generating core competencies. Integrated facilities management (IFM) services – are attracting large corporations and property owners seeking to cut operating costs, typically offering client savings of 15–20% on operation and occupancy costs. The high levels of expertise of facility service providers ensure demand from clients.

---

[3] Frost and Sullivan, *North American Integrated Facilities Management Services Market*, May 2003.

Companies are increasingly embracing IFM services as an effective way of minimising enterprise resources toward the operation and maintenance of critical building systems. Negotiating a good IFM services agreement, however, presents many challenges to both service providers as well as clients. For instance, one of the most important contractual issues involves reaching an agreement on disclaimers of consequential damage and limitations of liability. Strong, and mutually satisfactory resolution of such issues is necessary at the contract negotiation stage to avoid potential bitter experiences and unsatisfied customers.

Providing a bundled solution individually or through alliances is likely to be a major competitive factor. As companies pursue growth opportunities in international markets, the need for global real estate and IFM services such as corporate management and tenant representation is increasing, creating a demand for service providers that can provide such global services. The 'single-sourcing' trend is also stimulating demand for companies that can provide comprehensive services, best practices management, and consistent engineering standards.

International organisations are seeking global solutions in the delivery and standardisation of services, with early attempts from clients including HSBC, BP and Shell to reorganise and procure FM services on a global basis. Suppliers such as Johnson Controls and ISS have grown even more rapidly through mergers and acquisitions. Suppliers of multiple services are delivering services to organisations with global facility portfolios. Offshoring, as the term is also known, will bring further challenges for managing facilities, and the FM skillsets will be sought in regions that have not been considered for employment opportunities.

At the opposite end of the spectrum, FM also has to manage the foibles of those working and operating within the facilities. Aside from the general complaints of being too hot or too cold, there is a range of other complaints raised as part of the IFMA member 2003 Corporate Facility Monitor Survey on typical office complaints.[4]

In the survey 'It's too cold' and 'It's too hot' ranked one and two respectively, followed by, in order: (3) poor janitorial service; (4) not enough conference rooms; (5) not enough storage/filing space in workstation; (6) poor indoor air quality; (7) no privacy in workstation/office; (8) inadequate parking; (9) computer problems; and (10) noise level/too noisy.

Previous surveys conducted in 1991 and 1997 ranked the complaints of 'too hot' and 'too cold' first and second. All of the other complaints were ranked in different positions, but for the first time, the 2003 survey showed that the noise-level complaint had made the top ten. The most common complaint facility professionals reported hearing from upper management was related to the cost of facilities operations. Lack of space, the cleanliness and image of the facility, and the time required to complete construction and renovation projects were also cited.

---

[4] IFMA member 2003 Corporate Facility Monitor Survey – http://www.ifma.org

IFMA Member Survey 2003

Along with complaints about standard workplace issues from the IFMA Member Survey, a humorous list from employees emerged highlighting just how trying a facilities manager's job can be:

❏ I don't like the colour of the extension cord;
❏ The bathrooms are boring;
❏ People get stuck in the revolving door;
❏ The arms on my task chair are giving me breast cancer;
❏ The air in the building smells like bacon;
❏ Too much natural light;
❏ My workstation isn't located in a place that's going to get me a promotion;
❏ An employee did not want to move his complete Star Wars action figure set in order for housekeeping to clean his office;
❏ An employee's tie was caught in the deposit tray of the on-site ATM machine.

1.1.4   Global maturity of facilities management

A review of the UK FM market found the market worth £106.3 billion with annual growth levels expected to be between 2% and 3% in real terms from 2006 to 2010. IFM will grow 30% over the next five years, and from 8% to 9% as a proportion of the market. Contracted out ancillary services will reach £75.5 billion (at 2005 prices) by 2010, an overall increase of 21% compared to 2005. In-house services will remain static at around £36.5 billion.[5]

The FM market across Europe is currently in a strong growth phase with around 10%+ year on year, increases in revenues predicted to occur in the short to medium term, and current revenues in excess of $12 billion. Traditionally, organisations have employed in-house facility management teams to retain complete control of their operations, which is still prominent in continental Europe. This is compounded by a lack of awareness and misconceptions about what FM is and a certain level of resistance from strong labour unions.

In Australia, FM is one of the fastest growing and diversifying industries. With an annual turnover of more than AUD$60 billion, it is now one of Australia's major businesses sectors. The annual national investment of the FM industry contributes approximately 4% to GDP and there are more than 404 000 people working in FM service industries in Australia.[6]

More businesses are seeking to increase flexibility, which is perceived as an opportunity, that can be realised by outsourcing non-core activities. As a result

---

[5] *UK Facilities Management Market Development Report*, MBD Ltd, March 2006 – http://www.mbdltd.co.uk/UK-Market-Research-Reports/Facilities-Management.htm

[6] Facility Management Association of Australia Ltd – http://www.fma.com.au

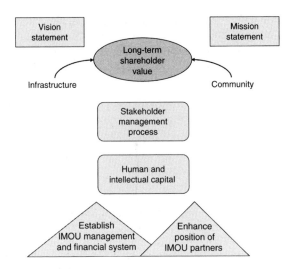

Figure 1.2 Global FM vision and mission model to deliver financial and stakeholder value.

of outsourcing, benefits can include cost reductions of up to 20% of existing costs and reduced headcount, enabling the business to maintain a focus upon core activities. In keeping with the predominance of in-house facility management teams is the provision of single-service providers for individual services. The growth in this sector is seen as providing businesses with the benefits of outsourcing in controlled areas where known benefits are available.

There are many approaches to outsourcing FM. The organisational structure employed to manage the resources is key to the successful day-to-day management and delivery of the services. Although it seems obvious, there is a huge variation in the types of structure employed not only based on the scope of services being managed, but also driven by the maturity of the FM market.

The FM outsourcing market is quite diverse in terms of its maturity, which is translated into the way in which organisations organise their resources. The development and growth of service delivery organisations are very much based upon the market conditions that exist globally. In this sense, across much of continental Europe there is a proliferation of maintenance based and cleaning based service-specific organisations. Progress is being made to standardise activities within Europe and globally. EuroFM is a liaison partner in the CEN standards: EN 15221 Facility management – terms and definitions and EN 15222 Facility management – guidance on how to prepare FM agreements.[7] The Global FM programme has integrated IFMA, BIFM and FMA Australia to pull together existing knowledge and information. Figure 1.2 provides a detailed mission and vision statement from the Global FM group and the mechanisms for implementation to deliver long-term stakeholder value.[8]

---

[7] Standardisation of FM activities in Europe – http://europa.eu.int/comm/enterprise/standards_policy/services/

[8] Global FM – http://www.globalfm.org

Table 1.1 Level of outsourcing maturity across Europe, North America and South-East Asia ranked as high (H), medium (M) or low (L)

Market	Outsourcing maturity (H, M, L)	Typical structures
UK	M/H	Matrix with increasing level of remote or shared service centres. Where single- or bundled-service sourcing is present, then a process-led functional structure is employed.
Germany and France	L	Traditional functional-service led structures predominate due to mainly single-service sourcing and the need to provide distinct and clear roles for staff.
Italy	M	Movement towards geographic structures for larger portfolios where total FM is employed, otherwise tends to be process-led functional.
Scandinavia	M	Matrix environments and remote teams becoming more common due to process outsourcing.
Benelux	L/M	Mainly process-led structures with service-led still prevalent.
Spain, Greece, Portugal	L	Mainly single-service structures in place due to an immature marketplace.
US	M/H	Matrix solutions with single- or bundled-service sourcing being driven by an increasingly process-led functional structure.
Australia	M	Matrix environments and remote teams becoming more common due to process outsourcing.
SE Asia	L	Mainly single-service structures in place due to an immature marketplace.

Generally speaking, the more mature or receptive the market is to outsourcing, the greater the move away from the traditional silo based functional teams, towards the more efficient matrix environment. Table 1.1 provides a summary of the outsourcing maturity across Europe, North America and South-East Asia which highlights the dramatic differences most obviously within Europe.

1.1.5 Future of facilities management

The role of FM is being cemented within organisations across the globe, with growth which has been documented in the United Kingdom and rolled-out in the other markets. There has already been movement with the expectation that in five years FM, HR and IT will be part of the same thing and will cease to exist separately. The industry will increasingly be seen as business infrastructure and facilities management leading towards being called infrastructure.

Of all the drivers for change, technology and communications rank as the most significant. The role of technology will play an increasing role in the provision of facilities – designs of buildings will be based upon wireless networks and integrated space requirements. The development of a co-ordinated communications system, capturing the building services, energy, life safety and security systems, is being investigated. Real-time and immediate reporting through websites not only serves to instantly engage people, but will undoubtedly change the way that work is performed and facilities managed in the future within and between organisations and suppliers. This change in technology usage will affect not only the physical parameters, but also the way the business is operated, the strategy and governance (Table 1.2).

The trend is certainly towards a greater level of e-business in the provision of FM services through a single common platform to provide a co-ordinated response to the procurement, management and delivery of services, materials and performance with computer aided facilities management (CAFM) systems. On-line

Table 1.2   Future of facilities management (adapted from article by Ron Adam and Nick Axford from *FM World*, 4 March 2004)

	From traditional shared services	**To workplace strategy**
Management	Linear, domestic organisations Bureaucratic, vertical leadership Rigid, command and control management Hierarchical decision-making	Complex, global networks Dynamic shared leadership Flexible management styles based on collaboration Centralised decision-making among distributed networks
Procurement	Internal sourcing – few partners Functional, silo approach to support services	Outsourcing – numerous partners Fluid, integrated workplace strategies
Governance	Limited accountability to shareholders Paper trails and hard copies; physical above virtual Corporate social responsibility not a high priority	More accountable to a broader range of stakeholders Technology an enabler and data-driving decision-making Corporate responsibility is a high priority
Strategy	Real estate experts Focus on physical and technical workers Loosely connected to core business goals	Workplace strategists Focus on highly skilled, strategic-knowledge workers Strongly connected and aligned with core business

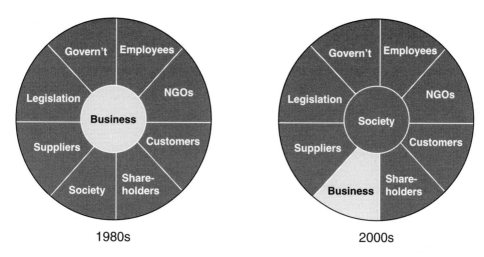

Figure 1.3   The increasing role of society in the operation of business.

booking systems for a range of services from meeting rooms, travel, stationery and hospitality will change the resourcing and delivery of services.

The move from the shared service delivery to a workplace strategy has incorporated the need for corporate responsibility to be taken as a core facet of FM. The Corporate Real Estate Network (Corenet) has included sustainability as one aspect of the Corporate Real Estate 2010 programme.

> By 2010, it is anticipated that most Global 1000 companies will use some form of triple-bottom-line reporting in place of traditional financial reports to key stakeholders. The adoption of this concept in turn will drive major changes in the way a company operates, particularly in the areas of facilities design, construction and management. Growing demand for sustainability as a design and construction element will influence real estate suppliers to stay competitive by delivering sustainability without the price premiums that currently exist. http://www2.corenetglobal.org/bookstore/index.vsplt has anticipated that sustainable buildings will deliver "cost-in-use" savings as the concept gains acceptance as a benefit rather than a drawback in facilities design and construction.[9]

Part of this move to a workplace strategy will see a greater role for the facilities within the community (Figure 1.3). Skills, and in particular, the shortage of skilled tradespeople is affecting the FM community across the globe. In part, this is through the rapidly changing requirements of the role of FM but also the limitations of attracting suitably qualified individuals into the profession. The demand for cost efficiencies and the need for operatives to act independently are limiting the potential for upskilling of new starters in the industry. IFMA are progressing the Chartered Facilities Manager (CFM) and BIFM are developing the Continual Professional Development (CPD) programme further.

---

[9] Corporate Real Estate Network – http://www.corenetglobal.org

The role of society will increase to take centre stage for the delivery of effective business operation and engagement. Increasing pressures from stakeholders will require the appreciation of societal needs and its inherent involvement on the perception of organisations affecting recruitment and retention, customer relations and ultimately the share price.

1.2 What is sustainable development?

Whilst the phrase and term sustainable development has gained popular momentum over the last ten years, the components of it date back many decades. From *Silent Spring* written by Rachel Carson[10] in the early 1960s describing a world affected by the chemicals and drive for increased productivity through to James Lovelock's Gaia philosophy[11] stating the role of 'mother earth', sustainable development has formed a spine through these books.

One of the first definitions of sustainable development was made in 'Our Common Future', the report of the Brundtland Commission, calling for development *'that meets the needs of the present without compromising the ability of future generations to meet their own needs'*.[12] Whilst still used today, over 500 definitions of sustainability and sustainable development have been spawned by various governments, professional bodies, institutions and organisations. A more commonly known terminology encompasses the environmental, social and economic principals captured as the 'triple bottom line'.

The UN Global Compact[13] has three principles devoted to the environment, demonstrating how important the environment has become: Precautionary, Proximity and Polluter Pays Principals. As far as the environment is concerned, it is not just a question of understanding cause and effect, but also the relationships between the different species. It is also a recognition of the fact that some changes, once they occur, are irreversible.

Sustainable development has been increasingly quoted over recent years as more organisations jump on the 'triple-bottom-line' bandwagon. The term has a variety of meanings dependent on the requirements of the organisations. This is largely as a result of a lack of guidance provided to fully define and capture what sustainable development means for specific sectors and activities. Whilst the provision of social, environmental and economic partnership is well provided for, the combinations and elements of the partnership are less well understood. This has meant a relatively unstructured process for organisations moving towards implementing sustainable development activities and processes into their day-to-day operations.

---

[10] Rachel Carson, *Silent Spring*, published by Boston Houghton Mifflin Company, (2002).

[11] James Lovelock's Gaia philosophy – http://www.ecolo.org/lovelock/whatis_Gaia.html

[12] Our Common Future, Report of the World Commission on Environment and Development 1987 (A/42/427).

[13] UN Global Compact – http://www.unglobalcompact.org

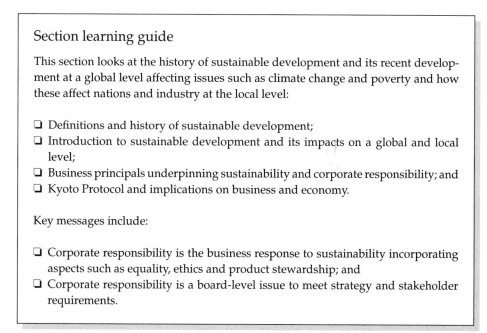

Section learning guide

This section looks at the history of sustainable development and its recent development at a global level affecting issues such as climate change and poverty and how these affect nations and industry at the local level:

❑ Definitions and history of sustainable development;
❑ Introduction to sustainable development and its impacts on a global and local level;
❑ Business principals underpinning sustainability and corporate responsibility; and
❑ Kyoto Protocol and implications on business and economy.

Key messages include:

❑ Corporate responsibility is the business response to sustainability incorporating aspects such as equality, ethics and product stewardship; and
❑ Corporate responsibility is a board-level issue to meet strategy and stakeholder requirements.

'As a world citizen, I believe that we have a responsibility to protect future generations from our actions today and that no-one – whether a nation or an industry – should stand on the sidelines.'

> Mike Clasper CEO,
> BAA plc CEPS 3rd Annual Brussels
> Climate Change Conference
> 19 April 2005

'There's enough on this planet for everyone's needs but not for everyone's greed.'

> Mahatma Gandhi

1.2.1   History of sustainable development

Sustainability as a concept has been in place for, certainly, well over 20 years, even dating back to the 1960s in the development of *Silent Spring* and arguably has been a concept in place throughout history from ancient civilisations. The recent development of corporate responsibility may or may not be a short-lived trend, but the underlying sentiments will certainly continue even if the name does not.

For organisations and businesses in the 1970s, many were looking to develop health and safety systems to comply with newly developed legislation, such as the Health and Safety Act 1974 in the United Kingdom. The development of environmental criteria took hold in the 1980s and 1990s with greater legislation and the extension of the Environmental Management System ISO14001 across organisations globally, leading to a greater customer and public awareness on the role business plays in the wider community.

In the last ten years the role of sustainability and corporate responsibility – the business response to sustainability – has become a pre-requisite for customers and investors. The United Nations Conference on Environment and Development (the Earth Summit) in Rio de Janeiro in 1992 and the Rio plus 10 conference in Johannesburg, South Africa, in 2002 have raised the profile and awareness of sustainability to a global audience. Delegates at Rio committed to Agenda 21, the Global Programme of Action on Sustainable Development.

Situation in the 1980s

- Many environmental and health and safety regulations governing many company activities;
- 'Environment' the main concern, largest growth area was environmental management systems;
- Large companies had policies, staff, documentation, performance management.

During the 1990s…business implications

- No longer sufficient to knowingly break the law – compliance was no longer a shield;
- Concept of sustainable development increases in status and credibility – balance between environment, social, economic activity;
- Range of unmapped additional areas – and role of 'stakeholders' complicated things further;
- Gap emerges between companies prepared to take action, and those needing further encouragement.

Into the 2000s

- 'Sustainable development' begins to morph into 'corporate responsibility' for business;
- 'Sustainable development' carried implication of holistic balance;
- 'Corporate responsibility' appears to include a range of issues with no specific priorities: business ethics, products stewardship, governance, globalisation and supply chains, human rights.

Sustainable development is therefore not a new practice or activity. It has developed over a period of time and will continue to change in the future.

1.2.2   What does sustainable development mean?

Sustainable development represents a process and a framework for redefining social progress and redirecting our economies to enable all people to meet their basic needs and improve their quality of life, while ensuring that the natural systems, resources and diversity upon which they depend are maintained and enhanced, both for their benefit and for that of future generations.

Economic	Environment	Society
Profitability, wages and benefits, resource use, labour productivity, job creation, human capital and expenditures on outsourcing	Impacts of processes, products, services on air, water, land, biodiversity, human health	Workplace health and safety, community relations, employee retention, labour practices, business ethics, human rights, working conditions

Figure 1.4 Typical issues and criteria comprising sustainable development.

Sustainability drives us to seek continuous improvements, in a way that integrates economic, environmental and social objectives into both our daily personal and business decisions and future planning activities (Figure 1.4). It also represents an approach for unlocking opportunities for improving sector competitiveness and enhancing reputation.

The UK Environment Agency published research in January 2004[14] that showed the link between social deprivation and environmental problems, and identified the need for a 'joined-up approach' to address environmental inequalities alongside social and economic problems. The research found that deprived communities suffer the worst air quality, and are more likely to live on tidal floodplains and near to polluting industrial sites. Although written for the UK market, the findings resonate with the global situation in both developed and developing countries.

Critical findings of the Environment Agency report included:

❏ In some parts of the country, deprived communities bear the greatest burden of poor air quality;
❏ Industrial sites are disproportionately located in deprived areas; and
❏ Tidal floodplain populations are strongly biased towards deprived communities.

At the opposite end of the spectrum, The Worldwatch Institute highlighting more than a quarter of the world's population has now entered the 'consumer class'. Higher levels of obesity and personal debt, chronic time shortages, and a degraded environment are all signs that excessive consumption is diminishing the quality of life of many people. Private consumption expenditure at the household level has increased fourfold since 1960. The 12% of the world's people living in North America and western Europe account for 60% of this consumption, while the one-third living in south Asia and sub-Saharan Africa account for only 3.2%.[15]

The report points out that in the US today:

❏ there are more private vehicles on the road than people licensed to drive them;
❏ the average size of refrigerators in households increased by 10% between 1972 and 2001 and the number per home rose as well; and

---

[14] Environment Agency Report: *Environmental Quality and Social Deprivation* (R&D Technical Report E2-067/1/TRA).

[15] The Worldwatch Institute Report, *State of the World 2004* – http://www.worldwatch.org

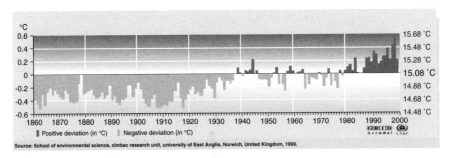

Figure 1.5   Trend in global average surface temperature. (Source: University of East Anglia.)

❏ new houses were 38% bigger in 2000 than in 1975, despite having fewer people in each household on average.

1.2.3   Global and local issues

Media interest has piqued public interest in a range of sustainable development issues, in particular, climatic change and poverty.

Climate change

The world's climate is changing and will change further in the coming decades as a result of the increasing concentration of greenhouse gases (GHGs), notably carbon dioxide, which accounts for 60% of the total impact of GHG emissions. Since the industrial revolution, levels of carbon dioxide in the atmosphere have grown by more than 30% as a result of burning fossil fuels, land-use change and other man-made emissions. This is amplifying the natural 'greenhouse effect' leading to global warming: the average surface temperature increased by 0.6°C during the twentieth century. The 2001 report of the Intergovernmental Panel on Climate Change (IPCC)[16] concluded that 'there is new and stronger evidence that most of the warming observed over the last 50 years is attributable to human activities' (Figure 1.5).

The period from the 1980s onwards has been estimated to be the warmest in the last 2000 years, and the ten hottest years on record have all occurred since the beginning of the 1990s. The IPCC estimates that global average surface temperatures could rise further by 1.4°C to 5.8°C by 2100. Rising emissions and a disrupted climate are leading to a range of impacts, such as more frequent heat waves, increased intensity of floods and droughts, as well as rising sea-levels.

These changes are already translating into real economic losses:

❏ In 2002, the severe floods across Europe generated direct losses of $16 billion;[17]

---

[16] Intergovernmental Panel on Climate Change – http://www.ipcc.ch

[17] *HM Treasury, Long-term global economic challenges and opportunities for Europe*, March 2005 – http://www.hm-treasury.gov.uk/documents/international_issues/int_global_index.cfm

❑ The 2003 heat-wave that affected much of Europe is estimated to have caused 26 000 premature deaths and an estimated economic cost of \$13.5 billion;[18]

❑ Hurricanes Katrina and Wilma in 2005 battered the Gulf Coast in the United States causing widespread damage and loss of life;

❑ Claims for storm and flood damages in the United Kingdom doubled over the period 1998–2003, compared with the previous five-year period, according to the Association of British Insurers;[19]

❑ Looking ahead, a 2.5°C rise in global temperature during this century could cost as much as 1.5–2.0% of global gross domestic product (GDP) in terms of future damage.[20]

It is widely believed that the gases emitted as we burn fossil fuels are the most likely reason for the past 25 years of warming. These gases contribute to the 'greenhouse effect' as they accumulate in the atmosphere, trapping the outgoing heat radiated from the surface of the earth. The earth has natural fluctuations in temperature and carbon dioxide levels, evidenced through investigation of ice cores from the poles. Temperature on the planet has increased over the past several thousand years since the last ice age. However, the role of mankind is exacerbating the natural variation in temperature through the addition of contaminants into the atmosphere faster than the planet can 'manage' itself and normalise the effects (Figure 1.6).[21]

The effects of climate change are going to cause a greater variability in weather patterns and an increase in major environmental events. Natural variability of climate due to solar output and volcanic eruptions partly explain the recent warming. Additional changes will result in droughts, floods, erosion and damage to buildings and roads.[22]

Use of resources

The increasing consumer demand is placing an increasing burden on the world's natural resources and habitats. The Worldwatch Institute has classified the 'consumer class', once limited to the rich nations of Europe and the United States and epitomised by large cars, plentiful diets and an abundance of waste, as now encompassing almost a quarter of humanity. Since 1960 the amount spent on household goods and services has increased fourfold, exceeding US\$20 trillion in 2000. However, the developing world is beginning to copy the trends of the richer

---

[18] *HM Treasury, Long-term global economic challenges and opportunities for Europe*, March 2005 – http://www.hm-treasury.gov.uk/documents/international_issues/int_global_index.cfm

[19] *A Changing Climate for Insurance*, 2004, Association of British Insurers http://www.abi.org.uk/ Display/File/Child/552/A_Changing_Climate_for_Insurance_2004.pdf

[20] IPCC Second Assessment Report, quoted in Commission of the European Communities, Winning the Battle against Global Climate Change, Communication, February 2005 – http://www.europa.eu/press_room/presspacks/climate/staff_work_paper_sec_2005_180_3.pdf

[21] *Climate Change – The UK Programme*, Department for the Environment, Food and Rural Affairs, UK Government 2006.

[22] Beacon Network Seminar on climate change adaption, Business in the Community 2004 – http://www.bitc.org.uk

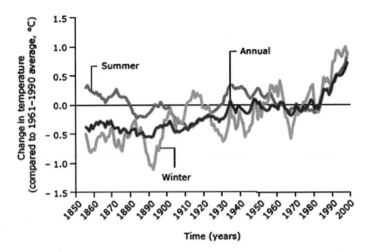

Figure 1.6   Changes in winter and summer temperatures over the past 150 years. (Source: EEA Signals 2004: A European Environment Agency update on selected issues.)

nations, with China and India now home to a larger consumer class than that in all of western Europe.

The rising consumption in the United States, other rich nations and many developing ones is more than the planet can bear. Forests, wetlands and other natural places are shrinking to make way for people and their homes, farms, shopping malls and factories.

World Living Beyond its Environmental Means – WWF[23]

The world is consuming 20% more natural resources a year than the planet can produce. Between 1970 and 2000, populations of marine and terrestrial species fell 30%. That of freshwater species declined 50%. 'This is a direct consequence of increasing human demand for food, fibre, energy and water ... humans consume 20% more natural resources than the earth can produce.'

What WWF calls the 'ecological footprint' – the amount of productive land needed on average worldwide to sustain one person, currently stands at 5.43 acres. But the earth has only 4.45 acres per head – based on the planet's estimated 11.3 billion hectares (27.9 billion acres) of productive land and sea space divided between its 6.1 billion people. The fastest growing component of the footprint is energy use, which had risen by 700% between 1961 and 2001.

North Americans are consuming resources at a particularly fast rate, with an ecological footprint that is twice as big as that of Europeans and seven times that of the average Asian or African.

---

[23] WWF Living Planet Report 2002 – http://www.wwf.org.uk/filelibrary/pdf/livingplanet 2002.pdf

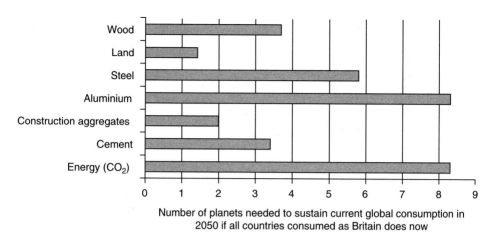

Figure 1.7 Ecological footprinting of commonly used materials based upon the number of planets required to support the UK (Source: Friends of the Earth).

Ecological footprinting has been developed as a mechanism to translate the use of resources and materials to enable easy comparison. The globe has an average footprint of 1.8 hectares per person based upon 6.1 billion people and 11.3 billion hectares of biologically productive space. Figure 1.7 shows the number of planets required to sustain the average person in the United Kingdom, highlighting the over-consumption of many widely used and commonly wasted materials. Each activity, whether at country, city or business level has an ecological footprint which can be compared to this average. Examples of ecological footprints from the European Environment Agency[24] include:

United States	9.7 ha/person
Canada	7.5 ha/person
Australia	7.0 ha/person
New Zealand	6.0 ha/person
United Kingdom	5.6 ha/person
China	1.6 ha/person

Fossil fuels are being depleted 100 000 times faster than they are formed; each person in the western economies consumes 60–80 tonnes of material each year on average, and uses roughly 11 tonnes of raw materials to produce one tonne of product.

---

[24] European Environment Agency – http://org.eea.eu.int/

The world's forests are still being destroyed mainly from the conversion of forests to agricultural land. The numbers measure net loss, taking into account forest growth from new planting and natural expansion. An average 7.3 million hectares was lost annually over the last 5 years. This was down from 8.9 million hectares (22 million acres) a year between 1990 and 2000.[25]

Waste

There are major differences globally in recycling levels. In 2002, almost 80% of municipal waste in England was sent to landfill sites, compared with around 50% in France and 7% in Switzerland. Likewise, England recycled approximately 20% of its municipal waste, while Germany recycled 52% and the Netherlands 47%.[26] The volumes of waste being generated are increasing from rising packaging materials and consumer led changes in products such as electronic goods.

Factors underpinning poor environmental performance on waste include the ready availability of cheap landfill sites, weaker regulatory controls and the absence of incentives for recycling, low public awareness and an inability or unwillingness on the part of many local authorities to invest in more expensive recycling and waste disposal options. Examples, such as Figure 1.8, are commonplace across the globe of general poor management of waste. In order to reduce the volume of waste generated, radical action is required. Where a fixed rate for waste management is paid, there is no economic incentive to reduce waste volumes, or to recycle and compost.

A variable rate encourages recycling so that good recyclers would be rewarded, while those who produce more rubbish would have to pick up the bill for their wastefulness. Such schemes are commonplace in countries with high recycling rates (e.g. Germany and Austria) where they have typically led to an increase of 30–40% in recycling and composting.

Recycling also has a number of adverse environmental impacts. For example, wastes have to be collected and transported to recycling facilities, which may be further away than landfill sites. They then have to be reprocessed into useful materials, and the relevant processes may be both relatively energy intensive and/or polluting. The only way to arrive at a clear answer as to whether recycling is environmentally beneficial is through a process called 'life-cycle analysis' (LCA), which seeks to measure all the environmental impacts of resource use, for both virgin and recycled materials, so that they can be compared validly.

Using the LCA methodology, it is clear for some materials, especially metals like aluminium and steel, that recycling has environmental benefits and is often cost-effective too, because the energy required for recycling is much less than for producing these materials out of virgin ore.

---

[25] UN Food and Agriculture Organisation – http://www.fao.org/

[26] Waste Indicators – http://themes.eea.eu.int/Environmental_issues/waste/indicators/

Figure 1.8   Typical example of waste disposal – litter being scattered in the open environment.

For glass, the energy balance tends also to be positive, but there can be problems with the product; recycling green glass, which dominates the waste glass stream, will not produce clear glass. Some of the benefits of recycling can be lost if the recycled glass is put to a different use (e.g. as aggregates in roads). This is still better than using raw materials and reduces the amount of waste being landfilled.

For paper and cardboard the environmental issues are more finely balanced, and depend on the distances the wastes have to be transported, on how the trees for virgin paper are grown, and whether they are replanted once cut down (Figure 1.9). Sometimes it is clear that the best use for waste paper and board in terms of energy would be for energy generation in incinerators, but this raises other environmental issues such as air emissions.

Surveys show that recycling is one of relatively few environmental actions that command widespread public support, probably because individual action can be seen to make a difference.[27] It could be counterproductive to reduce people's environmental commitment, in respect of recycling and perhaps impacting onto other issues, by drawing attention to the complexities of recycling's balance of benefits,

---

[27] Tim Jackson, *Policies for Sustainable Consumption; A report for the Sustainable Development Commission*, Centre for Environmental Strategy, University of Surrey and Laurie Michaelis, Environmental Change Institute, Oxford University.

LIVERPOOL JOHN MOORES UNIVERSITY
LEARNING SERVICES

Figure 1.9 Cardboard recycling.

or, even worse, to those relatively rare cases where recycling is not environmentally beneficial.

Pre-treatment of wastes as a means to reduce the volume of waste to landfill is becoming enshrined within legislation and corporate practice. The removal of hazardous materials, such as printed circuit boards from electronic equipment enables the rest of the waste to be treated as non-hazardous and become recycled, with only a small quantity deemed hazardous. Much of the pre-treatment takes place in parts of the world such as Gujarat State in India where labour is cheaper and therefore poor health and safety practices and low wages prevail.

Waste refrigerators are exported to developing countries, burnt in incinerators or find their way to landfill sites. The majority of fridges, built prior to 1996, contain harmful ozone depleting substances (ODS). In the European Union, electronic waste is the fastest growing waste stream, with the average EU citizen producing 17–20 kg of electrical waste every year, much of which contains potential pollutants. The vast majority of the waste in Europe still goes to landfills or incineration, despite general acceptance of the need for action to reduce inefficient use of resources and the risk of contaminants leaking into the surrounding soil, water or air, posing a risk to human health as well as the wider environment.

Aside from environmental issues, there are major commercial advantages to recycling including the recovery of valuable metals such as iron, steel, copper and aluminium from white goods, such as fridges, which typically consist of 70% metal. Once recovered, the metals can be sold back to the metals industries for recycling, to generate revenue. This not only produces economic but also environmental benefits. Recycling metals uses less energy. For instance, 20 times the amount of energy to produce one tonne of recycled aluminium is required to yield one tonne of new aluminium. As a consequence the need to locate and mine new metal ores is reduced, placing less pressure on the environment.

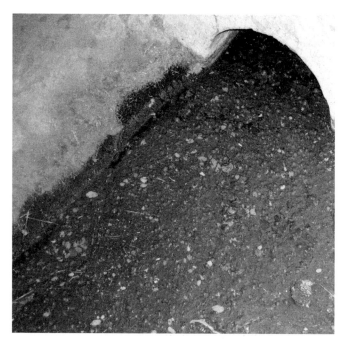

Figure 1.10   Water pollution is hindering the limited supply of available clean water.

Water

The earth is covered by 70% water, yet there is a known scarcity of water in the world. Flooding affects the south coast of the United States and Bangladesh on a regular basis, with water shortages causing conflicts between Egypt, Ethiopia and Sudan from the River Nile, and in the Middle East from the River Euphrates. Drinking water consumption patterns represent the availability of water: United States consumes 590 litres per person per day; France consumes 290; China 88 and Mali 12.[28]

Pollution of existing water systems leads to diseases which can kill people and limits supplies, hindering development (Figure 1.10). In 2002, 2.6 billion people were without improved drinking water – over 40% of the global population (Figure 1.11).[29]

❑ 5 litres of oil poured into a lake can spread to cover an area the size of two football pitches;

❑ Most pollution incidents are the result of ignorance, apathy or neglect; and

❑ Just 1 litre of solvent is enough to contaminate 100 million litres of drinking water.

---

[28] UN Food and Agriculture Organisation – http://www.fao.org/

[29] UK Environment Agency – http://www.environment-agency.gov.uk

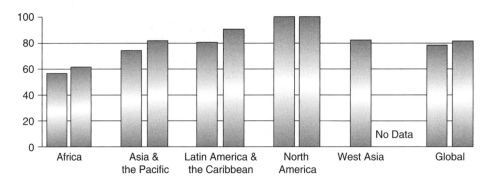

Figure 1.11 Trends with the access to clean drinking water (% of total) by region, 1990 and 2002. (Source: GEO Data Portal, adapted from WHO/UNICEF 2004.)

Biodiversity

There are many issues affecting biodiversity – but perhaps the two greatest issues are the loss of habitat due to forest destruction and oceans. Recent drives to improve the level of biodiversity such as small-scale forestry have resulted in non-indigenous planting causing greater damage through the introduction of species unnatural to the local food chain.

Ten million hectares of ancient forests are being cleared or destroyed every year – the equivalent to an area the size of a soccer pitch every two seconds. Whilst there are around 1350 million hectares of ancient forests remaining undisturbed, this represents only 7% of the earth's land surface, and only one-fifth of the forests' original size. Since 1950, 20% of the world's ancient forests have been cleared, with those in Indonesia and central Africa likely to have gone in a few decades if forest destruction continues at its present pace.[30]

The primary causes of forest loss and degradation varies, but revolves around agricultural expansion, mining, settlement, shifting agriculture, plantation establishment and infrastructural development. These ancient forests are home to millions of forest people who depend on them for their survival – both physically and spiritually. It is estimated that some 1.6 billion people worldwide depend on forests for their livelihood and 60 million indigenous peoples depend on forests for their subsistence. Forests also house around two-thirds of the world's land-based species of plants and animals. That's hundreds of thousands of different plants and animals, and literally millions of insects – whose futures also depend on the ancient forests.[31]

The world's ancient forests maintain environmental systems that are essential for life on earth. They influence weather by controlling rainfall and evaporation of water from soil. They help stabilise the world's climate by storing large amounts of carbon that would otherwise contribute to climate change. Over the past 8000 years, they have faced a dramatic reduction (Figure 1.12). Many scientists believe that the world is facing the largest wave of extinctions since the disappearance of the

---

[30] Rainforest Action Network – http://www.ran.org

[31] Greenpeace – http://www.greenpeace.org

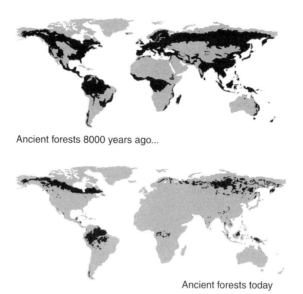

Ancient forests 8000 years ago...

Ancient forests today

Figure 1.12   Disappearance of native ancient forests over the past 8000 years. (Source: World Resources Institute, Washington 1997, revised by Greenpeace 2002.)

dinosaurs 65 million years ago. It is estimated that nearly 24% of mammals, 12% of birds and almost 14% of plants face extinction. Most of these extinctions will be due to habitat destruction; most of these habitats are ancient forests.

Recent research by the World Resources Institute (WRI) concludes that 'commercial logging poses by far the greatest danger to frontier forests ... affecting more than 70% of the world's threatened frontiers'.[32] The way the logging industry operates also exacerbates the problem. Unplanned tree cutting and inefficient processing leads to an enormous wastage of wood, while lack of transparency within the industry makes it very difficult to trace the exact source of wood supply. Illegal logging is adding to this problem accounting for up to 90% of the logging that takes place from Indonesia, 60–80% in the Brazilian Amazon and 50% in Cameroon between 1999 and 2004.

The oceans cover two-thirds of the planet, host 80% of life from microscopic plankton to great whales and provide half the oxygen the planet needs. The increasing exploitation of marine life coupled with the burning of fossil fuels and the development and dumping of chemicals into the oceans outpace their ability to cleanse themselves and maintain a natural balance.[33] Commercial fishing and tourism is placing a strain on the oceans to replenish themselves:

❏ Industrial fishing fleets are far exceeding the ocean's ability to recover, with smaller fish sought as larger fish are wiped out. It is estimated that 90% of the world's large fish are already gone;

---

[32] World Resources Institute – http://www.globalforestwatch.org

[33] Greenpeace – http://www.greenpeace.org

LIVERPOOL JOHN MOORES UNIVERSITY
LEARNING SERVICES

❏ Every year, up to 300 000 whales, dolphins and porpoises die in nets and 100 000 albatrosses are caught on hooked fishing lines. Turtles, seals and sharks are also victims of indiscriminate fishing practices;

❏ The extension of fishing grounds into the waters of Africa and the Pacific is increasing the conflict between countries over fishing rights;

❏ In 1998, 16% of the world's corals were severely damaged and in south Asia and the Indian Ocean, half of the reefs lost the majority of their coral; and

❏ Chemicals, oil, plastics, sewage, industrial discharges, intentional dumping, mining, general rubbish and tanker incidents are causing severe marine pollution.

Supply chain

Despite the economic benefits associated with increases in trade worldwide, inequalities between rich and poor are widening, both between and within countries. According to Oxfam,[34] with only 14% of the world's population, high-income countries account for 75% of global GDP, which is approximately the same share as in 1990. According to the 2001 Human Development Report,[35] 1.2 billion people worldwide live on less than $1 a day, and 2.8 billion on less than $2 a day. Of the 4.6 billion people living in developing countries, more than 850 million are illiterate, nearly a billion lack access to improved water resources, and 2.4 billion lack access to basic sanitation.

An increasing volume of trade being exported to lower labour cost locations is exacerbating the gaps described. Electronic goods and clothing are typical examples of goods sourced in low-cost markets, providing a direct link between day-to-day purchases and the issues related to pay and working conditions.

Inequalities in trade are reinforcing these wider inequalities: for every $1 generated through exports in the international trading system, low-income countries account for only 3 cents. Export success in developing countries has also been highly concentrated. East Asia accounts for more than three-quarters of manufactured exports, and an even larger share of high-technology products.[36] Sub-Saharan Africa, on the contrary, has suffered from a dramatic decline in its share of world exports, which currently only accounts for 1.3% of exports of goods and services. This has resulted in an enormous decline in living standards and a great increase in poverty.

Declining prices of commodities such as cocoa, coffee, minerals and metals are not a new phenomenon. Unfortunately, many of the world's poorest people depend heavily on primary commodities. More than 50 developing countries depend on three or fewer such commodities for more than half of their export earnings. While low prices may seem attractive for multi-nationals, some businesses have started to

---

[34] Oxfam – http://www.oxfam.org

[35] UNDP Human Development Report 2001 – http://hdr.undp.org/reports/global/2001/en/

[36] Trade Inequalities – http://www.maketradefair.com

Figure 1.13   Images of poverty standing next to areas of prosperity are commonplace.

consider different purchasing operations whereby they get involved in long-term contract arrangements with farmers.

Direct procurement in coffee and cocoa, for instance, involves the roaster buying directly from the farmer so that the farmer gets the full value from his work. Similar to the Fairtrade methodology, this not only allows the farmer to receive a higher price for his coffee, but also pushes him to expand his farming potential. It also induces quality improvement as payment is based on quality criteria. Other companies have joined international initiatives that attempt to tackle some of the negative aspects associated with commodity prices. Industry-led initiatives such as the Sustainable Agriculture Initiative of the food industry, established by Danone, Nestlé and Unilever, are innovative business-to-business ways for developing and communicating worldwide the challenge of sustainable agriculture and involving different stakeholders.

Poverty

The issue of poverty has been made public during the mid-1980s following the appalling images from the famine in Ethiopia and other African countries. Images of poverty are commonplace, not only in developing countries as commonly perceived (Figure 1.13) but also within many cities including London, New York, Los Angeles and Berlin.

In recent years, the question of poverty has been related to the reduction of debt repayments made by the developing nations. These debt repayments result from loans provided by the wealthier nations which require repayment. However, the repayments constitute a significant level to some developing nations' affordable revenue.

The UN Millennium Development Goals have highlighted the following:[37]

❑ Worldwide, over 1 billion people live below the extreme poverty line of $1 a day, including 21% of the developing world's population;
❑ Every 3.6 seconds a person dies of extreme poverty – many of whom are children;
❑ Every 30 seconds an African child dies of malaria; and
❑ 68 000 women die each year from unsafe abortions.

United Nations' Millennium Development Goals

Goal 1: Eradicate extreme poverty and hunger
Goal 2: Achieve universal primary education
Goal 3: Promote gender equality and empower women
Goal 4: Reduce child mortality
Goal 5: Improve maternal health
Goal 6: Combat HIV and AIDS, malaria and other diseases
Goal 7: Ensure environmental sustainability
Goal 8: Develop a global partnership for development

Human rights

In 2003, the United Nations launched an initiative to clarify the responsibility of business in promoting and protecting human rights and to develop principles that could lead to the investigation and censure of businesses that failed to meet their human rights obligations. A number of countries including China, Turkey and Yemen have been highlighted for their poor human rights' records by groups such as Amnesty International.

Seven companies, ABB, Barclays and Novartis, MTV Europe, National Grid, Novo Nordisk and The Body Shop, have drawn up a list of actions for businesses to take to address issues raised by the UN. 'Minimum' actions include compliance with national laws and regulations. 'Expected' actions cover such issues as whistle-blower protection and payment of a living wage, while 'desirable' actions include HIV/AIDS treatment programmes and withdrawal from countries where forced labour is prevalent.

The principles will pull together existing international standards on labour, health, environmental and discrimination issues, with the intention that companies should be subject to the kind of enforcement by the UN Commission for Human Rights that has previously applied only to nation states.

---

[37] UN Millennium Development Goals – http://www.un.org/milleniumgoals/

Figure 1.14   Air pollution is not always as visual smog.

Air pollution

A study based at the University of Southern California[38] tracked the health of almost 23 000 people from 260 Los Angeles neighbourhoods and found the death toll from fine particles could be up to three times greater than previously thought. For each increase of 10 $\mu g/m^3$ of fine particles in a neighbourhood's air, the risk of death rose by 11–17%. Fine particle levels can differ by about 20 $\mu g/m^3$ from the cleanest parts of Los Angeles to the most polluted, leading to huge differences in life expectancy.

Microscopic particles less than 2.5 μm ($PM_{2.5}$) in diameter pose the greatest problems to health because they can penetrate deep into the lungs and sometimes even enter the bloodstream. Particulates this small are often found in smoke, vehicle exhaust, industrial emissions and haze, driven by the burning of fossil fuels. Scientists also tracked ozone pollution, but found no link between ozone levels and mortality.

The particulate matter, generated from industry and specifically from vehicle emissions, is a common occurrence throughout major urban environments – smog affects Athens, Mexico City and Mumbai causing hazy skies (Figure 1.14).

1.2.4   Kyoto Protocol and emissions trading

In June 1992, the governments of the world signed the UN Framework Convention on Climate Change (UNFCCC) to combat the growing threat from climate change. Its ultimate objective was 'to achieve ... stabilisation of greenhouse gas concentrations in the atmosphere at a level which would prevent dangerous anthropogenic interference with the climate system'.[39] Subsequently, the Kyoto Protocol

---

[38] University of Southern California Air Pollution Report – http://www.usc.edu/uscnews/stories/11608.html

[39] UN Framework Convention on Climate Change – http://unfccc.int/

was adopted by 159 countries in 1997 in Kyoto, Japan. Among the 117 countries that have ratified the Protocol since, 40 of them are industrialised nations that agreed to reduce their emissions of six key greenhouse gases by 5.2% from their 1990 levels by 2012. The United States bailed out on the Kyoto Protocol in 2001, with President W. Bush saying that the treaty would place too heavy a burden on the US economy.

Kyoto came into force in February 2005 as a binding agreement when Russia ratified the treaty taking the level beyond the 55% of 1990 developed-country emissions. It has been ratified by almost all of Europe, Japan, Canada, Russia and New Zealand, collectively accounting for 61.4% of the required total. The United States (36.2%) and Australia (2.1%) have said they will not ratify. The Kyoto Protocol provides an opportunity to modernise technologies which many companies, US and Australian included, don't want to fall behind. It is estimated that the financial value of the European carbon market will be Euro 10 billion (£6.9 billion) a year in 2007.[40]

In 2001, negotiators agreed that signatories will face mandatory consequences if they fail to meet their targets. Countries that miss their emission targets would be excluded from 'emissions trading', buying and selling the right to pollute, while a panel to be set up by member governments would address alleged violations. The Protocol also foresees emission credits for forests that soak up carbon.

As a result, the European Union has launched an emissions trading system under which European companies that emit less carbon dioxide than permitted can sell unused allotments to those who overshoot the target. The motive for profit is expected to drive efforts and technology and bring 'substantial cuts' in emissions of carbon dioxide. The European system foresees fines for companies that exceed their emission limits without managing to trade. The EU scheme has created a new market in carbon dioxide allowances valued at around Euro 35 billion per year, rising to over Euro 50 billion per year at the end of the decade.[41]

Commitments to the Kyoto Protocol could be met through the purchasing of carbon credits and forest sink credits, without reducing emissions of greenhouse gases at all. Since the 1990 baseline for Kyoto obligations, many smokestacks in the former Soviet Union have gone cold, enabling Russia and the other successor countries to sell their carbon credits for the consequent drop in emissions. Forest sink credits arise under a Kyoto provision which recognises that plantation forests, established on land not previously forested, are withdrawing carbon dioxide from the atmosphere while they are growing. New Zealand expects to have more forest sink credits than it would need to cover the 33% increase in its emissions since 1990.

The United States, Australia, Canada, India, Japan and South Korea signed a clean technology deal in summer 2005. Whilst many of the countries have agreed to support the Kyoto Protocol, both the United States and Australia have implemented the deal as a direct alternative. The United States is investing $5 billion per year on the development of technologies to support a reduction in carbon

[40] Kyoto Agreement, European Programme – http://europa.eu.int/comm/environment/climat/kyoto.htm

[41] EU Emissions Trading Scheme – http://europa.eu.int/comm/environment/climat/emission.htm

intensity – the largest such programme in the world. The United States has also adopted the target of reducing emissions over the next ten years not in absolute terms but relative to economic output, and promised more action in 2012 if that goal is not met 'and the science warrants it'.[42] Nine north-eastern US states are introducing mandatory emissions control measures for industry. The agreement sees the states (Connecticut, Delaware, Maine, Massachusetts, New Hampshire, New Jersey, New York, Rhode Island and Vermont) cutting emissions from large power stations by 10% by 2020.[43]

In the United Kingdom, the Carbon Trust has studied the implications of climate change for brand value, and concluded that it could become a mainstream consumer issue by 2010, placing considerable value at risk for a range of sectors. Finally, investors have come together to seek enhanced disclosure from companies. At the global level, the Carbon Disclosure Project has now mobilised 143 institutional investors with over $20 trillion under management to improve disclosure from the world's leading 500 companies.[44]

Making progress on carbon management[45]

BP: In 1997, BP set a target of reducing GHG emissions from its own facilities by 10% from 1990 levels by 2010. It achieved this goal nine years early at the end of 2001, saving $650 million in net present value. A new target, ensuring that net emissions do not increase between 2001 and 2012, has been set.

BT: Since 1991, BT has managed to cut energy-related CO_2 emissions by 80% – equivalent to 1.5 million tonnes a year. In October 2004, BT announced the world's largest purchase of green electricity, reducing emissions by a further 325 000 tonnes a year. The company's target is to cap 2010 emissions at 25% below 1996 levels.

HSBC: In December 2004, HSBC committed itself to become the world's first carbon neutral bank through a combination of enhanced efficiency, purchasing energy from renewable sources and offsetting the remaining CO_2 by investing in carbon credit or allowance projects, which was achieved in 2006.

Rio Tinto: In 2000, Rio Tinto launched a climate change programme, which includes a number of commitments, such as a 4% reduction in GHGs per tonne of product between 2003 and 2008. It is also working with customers to improve the use of its products, notably talc and coal.

---

[42] US Emissions Programme – http://en.wikipedia.org/wiki/Kyoto_Protocol

[43] Emissions reduction programme by US States – http://www.ens-newswire.com

[44] Carbon Disclosure Project – http://www.cdproject.net/

[45] *The Carbon 100: Quantifying the Carbon Emissions, Intensities and Exposures of the FTSE 100*, Trucost June 2005 – http://www.trucost.com

One of Kyoto's limitations is that it does not place any constraints on developing countries, whose collective emissions of greenhouse gases are estimated to overtake those of developed countries within 20 years. The rationale is that the accumulated emissions of industrialised countries over the past century-and-a-half is doing the damage and rich countries are better placed to incur the costs of reducing emissions than countries still struggling to escape from poverty. Despite this, a number of developing countries, such as Mexico, China, Brazil and India, have managed to cut their emissions quite significantly from the transport, energy and forestry sectors, while emissions from the United States have risen year on year.

The Kyoto Protocol can only be a 'first step' toward negotiating deeper cuts in greenhouse gas emissions primarily because it sets very low targets compared to what scientists say is necessary in order to keep climate change under control. Recent rounds of international climate talks have already included negotiations on greenhouse gas emissions after 2012 to include many developing countries currently excluded.

1.2.5 Corporate responsibility

Companies are under increasing pressure from key stakeholders to be transparent about their values, principles and performance regarding sustainable development. One response to this pressure is the increase in 'sustainability reporting'. It is quite clear, however, that reporting is only the tip of the iceberg. Companies will find it difficult to continue to produce relevant and reliable reports without having internal management and information systems that support this undertaking. The key challenge is to integrate sustainable development issues into mainstream business processes and systems to determine how well companies implement their policies into practice.

The role of organisations in the global economy is increasingly coming under the scrutiny of investors, analysts and pressure groups. Activities such as the riots seen at the G8 summits are no longer greeted by public scorn, but with questions about why governments and private organisations are not pulling their weight to support a sustainable global economy (Figure 1.15).

There are a number of underlying principles for organisations to deliver sustainability:

❑ Sustainability must be the organising principle incorporated as part of the central business plan and processes;
❑ Natural resources and systems have finite limits and all economic activity must be constrained within those limits;
❑ Economics must ensure basic needs are met and maintained equally across the globe;
❑ Cost of pollution must be internalised – captured as part of the life-cycle costing and decision-making process;
❑ There is no blueprint – off-the-shelf solutions will not be appropriate or applicable to align with the business strategy; and

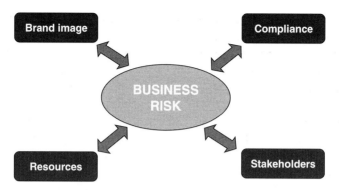

Figure 1.15 Relationship between business risks and activities which CSR (Corporate Social Responsibility) and SD support.

❏ Lack of scientific certainty should not be used as an excuse to delay taking cost-effective measures to prevent damage.

Multi-nationals have been the subject of high-profile consumer boycotts in recent years. An increasing number of consumers switch or boycott brands for ethical reasons, with a significant number of consumers refusing to return to the product subsequently.[46] Food, household appliances, cosmetics and tourism were among the most frequent choices of purchases for ethically minded consumers. This is coupled with an increased demand by consumers for multi-national companies to meet the highest corporate responsibility standards. Below are a few examples of companies and products that have been the subject of widespread boycotts.

Nestlé: For the past 20 years, the Swiss-based multi-national has been condemned for promoting powdered breast-milk substitutes in the third world, which critics claim contributes to the death of babies; a recent demand that Ethiopia repay its £3.75 million debts was retracted; and contaminated children's food taken off shelves across Europe.

Gap: The anti-Gap movement was galvanised when a demonstrator set a pair of Gap trousers alight at the World Trade Organisation conference in Seattle in protest against working conditions at factories in Cambodia, Indonesia, Bangladesh and Mexico. Since then, Gap have incorporated stringent checks on factories and suppliers, disclosing progress through their annual reporting.

Esso: The oil giant and its Texas-based parent company ExxonMobil have long been targeted over their environmental policies and alleged funding of the US President George Bush's election campaign.

Shell: The oil giant's conversion resulted from two disastrous events in 1995. Almost simultaneously Shell stations in Germany were attacked over the company's plans to sink the Brent Spar oil rig in the North Sea, and Shell was blamed for

---

[46] Co-operative Bank's Ethical Purchasing Index – http://www.greenconsumerguide.com/epi.php

the execution of Nigerian author Ken Saro-Wiwa, who had protested against Shell's environmental record in his country. To understand how it had so badly managed its public image, Shell commissioned a global survey of society's expectations. Since 1997 in its annual report, Shell has recorded its polluting emissions, its energy efficiency, and a range of other statistics to quantify its environmental and social performance. Shell decided to introduce two phrases: firstly, 'contribute to sustainable development' and second 'support for fundamental human rights'. However, taking this path has led to difficult questions about the potential clash between social and environmental concerns and short-term profitability. Shell took the decision never to drill on world heritage site land – a bold step since it was having more trouble than many rivals in finding new oil reserves.

The approach of corporate responsibility (CR) takes a similar approach to sustainable development, though it lays a greater emphasis on global activities and responsibility to those in the third world through the procurement or manufacturing of goods. CR is also broader in its context, taking into account elements of health and safety, ethical responsibility and investment in personnel.

Kofi Annan, Secretary General of the United Nations, proposed in 1999 for a global commitment by business to make international trade 'work for all the world's people'. The solution was to harness the increasing business trend towards 'corporate responsibility' and help companies develop a global, values-based system of management rooted in internationally accepted principles. This resulted in the launch of the UN's Global Compact in July 2000 to which hundreds of companies have signed up representing virtually all industry sectors on every continent alongside a range of other standards: The UN Global Compact principles, the OECD guidelines for multi-national companies, the International Labour Organisation (ILO) standards and declarations, the Universal Declaration of Human Rights, the Global Reporting Initiative (GRI) sustainability reporting guidelines, the Global Sullivan Principles, the Social Accountability 8000 (SA 8000) international workplace standard.

Figure 1.16 describes sustainability based upon the size of service area, from local to global, and the values of the organisation from capital to social. Mapping onto this diagram, many global organisations are driven by shareholders to deliver increasing returns, expansion into new markets and reduce baseline costs, placing them squarely within the capital and globalisation sectors. Alternatively, many small and medium size enterprises (SMEs) are more community based, but still are driven by a combination of maintaining community relations alongside generating profits. A common message covering all bodies is difficult due to the mixed drivers affecting them.

While critics of CR have said that it is not part of business core purpose, i.e. to do business and make a profit, many companies no longer agree. The market is changing, and CR is becoming a vital part of staying competitive, retaining talented staff, and satisfying customers' expectations. Brand-name companies like Shell, Nike and Nestlé have discovered through high-profile scandals concerning the environment, human rights, health and labour conditions that they have to take society's concerns seriously in order to preserve their licence to operate.

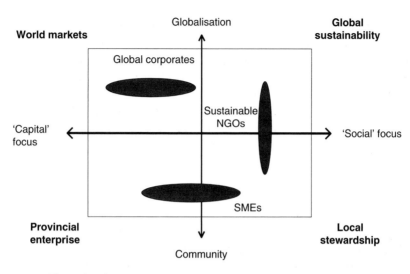

Figure 1.16   The role of various bodies in the modern marketplace.

Increased scrutiny via the internet also means organisations have to be rigorous in the way they address these issues. 'Greenwashing' or covering up their activities with symbolic gestures, will only have limited effect before stakeholders, including employees, demand greater transparency and more action. The next wave will see CR consolidated and integrated as a core business strategy. Companies that are left behind will be at a distinct disadvantage in the coming decade as CR becomes firmly embedded worldwide.

At the first world summit on sustainable development in 1992, companies were seen as the bad guys and were subject to attack by a wide range of stakeholders, including NGOs and civil society groups. The small delegation of business leaders that attended the meeting saw that something needed to change in business' approach to this debate. They went on to form the World Business Council for Sustainable Development, which comprises around 180 major multi-nationals.

The world's largest corporations now control about 25% of the world's economic output. For several years, sales of large groups like General Motors and Royal Dutch Shell have outnumbered the GDP of countries like Malaysia, the Philippines and Venezuela. Increased influence by multi-national enterprises through their purchasing, production and investment decisions and their development of new technologies has also produced a demand for increased accountability for their impact on society and the environment. This push towards greater responsibility has been driven by a variety of stakeholders, including national governments, consumers, non-governmental organisations and investors. However, an important feature of the growing range of CR initiatives by businesses is that they are predominantly of a voluntary nature and based on self-regulation by companies.

In response to increased importance of CR and pressures by the investor community, the number and importance of specialised ethical investment funds has increased significantly. In January 2002, 279 green, social and ethical funds operated in Europe. From January 2000 until the second quarter of 2001, ethical financial products available to the private clientele had increased by 58%.[47] Moreover, the global demand for ethical investing opportunities is growing.

Ultimately, businesses are required to operate within the legislative and economic framework created by governments. However, increasing globalisation is posing significant challenges that require new thinking about global governance, in particular the lack of integration of developing countries into the globalising world economy and the reduction of developed country protection and subsidies. Recognising the role that public policies can play in helping to globalise CR, the European Union has recently started to invite business representatives, international organisations and other partners to discuss how today's widely shared concerns around governance can be tackled. In a dialogue with business and civil society, the commission is discussing the implementation of innovative tools such as sustainability impact assessments in the coming trade agenda, and 'soft law' policy instruments as complements to hard law, including benchmarking and peer review, non-hierarchical governance, and co-regulation.[48]

At the 'micro' or national level, aligning business practices in host countries with corporate ethics represents a completely different challenge. Many companies face the reality of operating in countries with repressive regimes, where the rule of law is weak or absent, and where the independence of the judiciary is questionable. Positive case studies point out, however, that even in these conditions businesses can make a difference. For example, Reebok has encouraged freedom of association in countries like Indonesia by implementing innovative communication systems through which workers can express workplace concerns. The Pentland Group, faced with the option of pulling out of Sialkot in Pakistan, decided to stay in the area and succeeded in implementing a system to tackle child labour.

However, CR will only make a visible difference if the concept is fully integrated into corporate principles and practices, and if progress is monitored over time. Third-party involvement in this process is likely to improve credibility. The recent sustainability assessment, *Ten years after Rio*, published by the UN Environment Programme (UNEP) in 2002,[49] emphasises the need to integrate environmental and social criteria into mainstream business decision-making, while at the same time improving the implementation and monitoring of voluntary initiatives and self-regulation. As the world's markets and societies become increasingly globalised, companies will be called on to be more decisive in their responsibilities to society and the environment. However, the evidence thus far indicates that all parties stand to benefit from this new way of doing business.

Whilst the concept of CR has embedded itself within multi-nationals, the culture and processes of small and medium size enterprises (SMEs) have hardly changed at all – especially those at the smaller end of the spectrum. Small business owners

---

[47] Sustainable Investment Research International, January 2002 – http://www.siricompany.com

[48] EU CSR Programme – http://www.eu.int/comm/enterprise/csr/index.htm

[49] *Ten Years After Rio: Preparing for the World Summit on Sustainable Development in 2002*, published by the UN Environment Programme (UNEP) – http://ec.europa.eu/environment/agend21/index.htm

show little awareness and lack of interest due to a lack of resources, skilled staff and technical expertise, poor access to finance and often fall victim to burdensome red tape and regulation. Whilst this leads to SMEs being a significant source of environmental pollution, given the number and range of organisations, many have integrated links with the local community, cultural and religious environment, providing a strong social focus for employment. The role of larger organisations is to support the SMEs and, through the supply chain, build the skills and standards for the smaller business and local community. Responsible companies can help by engaging with local businesses and establishing small or micro enterprises that provide services that they can purchase.

1.3   Relationship between FM and sustainability

The incorporation of sustainable development in the day-to-day functions of facilities management is not new, and has been met, to some degree, as part of the function itself. The FM industry employs some of the lowest paid staff and manages environmental hazards including those related to waste and utilities. Organisations are moving towards the next level of proactive engagement of sustainability linked to the provision and delivery of services.

There exists a large volume of information available and organisations involved in the delivery of sustainability within FM, from website information to trade articles pitched at varying levels. Appendix 1 provides a collation of some internet based sources of information found to be useful.

Section learning guide

This section links the concepts of sustainable development and facilities management together and describes what sustainable FM is. It specifically covers:

❑ The role of FM in sustainable development;
❑ The requirements of the facilities manager;
❑ Global alliance for building sustainability; and
❑ Description of the 12 issues of sustainable FM.

Key messages include:

❑ FM has a critical role in supporting business to deliver corporate responsibility at a strategic and operational level; and
❑ Involvement of FM at the decision-making stage will deliver more cost-effective solutions.

'Our biggest challenge is to take sustainable development, an idea that seems abstract, and turn it into a daily reality for all the world's people.'

Kofi Annan, UN Secretary General

1.3.1   Sustainable development and facilities management

Sustainable development is becoming increasingly important to organisations through the development and implementation of sustainable policies, which translates the strategy into action and demonstrable evidence of progress. Increasing numbers of companies throughout the building supply chain, including investors, property owners and service delivery companies, are developing and implementing policies. The unresolved challenge is to co-ordinate the policies and activities of the parties in this chain so that each can demonstrate that its policy is being observed and its objectives met.

There is a major role that the FM industry can play, by influencing colleagues and the management hierarchy within the clients' organisation to understand the benefits and impacts of sustainable development. This will impact upon service delivery to ensure they incorporate sustainable criteria, such as energy reduction, waste minimisation, procurement controls and fair pay. Together, these have a balancing effect on the budget with net savings achievable over the lifetime of a contract whilst providing improved service delivery. FM now has a strategic role to play within business, utilising property performance metrics to support the decision-making process.

Since the 1990s, environmental management has gone from an activity restricted to the few 'tree-huggers' to a major business, and it is a key requirement for professionals involved in the design, construction and management of any property. The environment has become a key news item, with stories reaching front page headline status highlighting the focus it is maintaining with the public. Employees and sub-contractors are increasingly asking questions within the workplace about environmental practices, and the commitment to waste minimisation and utility efficiency.

The facilities manager has had to take into account an array of legislative requirements to manage the building, and additional pressures of meeting external certification or stakeholder requirements. The building is at the forefront of an organisation's persona, portraying its values and beliefs.

This has meant the role of the facilities manager has grown markedly to encompass activities such as waste minimisation, recycling initiatives, energy management and utility reduction to meet their customers' and clients' expectations.

1.3.2   Global Alliance for Building Sustainably (GABS) charter

The GABS charter was developed in 2002 from a growing recognition by the United Nations of the impacts the built environment has on global issues including waste, resources and climate change. The inclusion of a range of functions involved in the scope of the built environment provided recognition of the complexity in the management of facilities and the co-operation of FM with the other functions to deliver the changes required.

The inaugural meeting of the GABS, held in Johannesburg, South Africa, on 29 and 30 August 2002, was one of the first partnerships to emerge from the world summit on sustainable development (WSSD). These partnerships are UN-endorsed action-oriented coalitions, focused on deliverables that would contribute in translating political commitments into action partnerships and initiatives to implement Agenda 21.[50]

Proposed action plan

Let's get together to:

❑ understand what is happening in the area of sustainable building and construction and map out the key players;
❑ develop synergistic networks;
❑ discuss long-term goals for the development of sustainable building and construction.

It is the hope of both the World Resources Institute and the United Nations Environment Programme that this meeting will stimulate thinking on how we can collectively move towards the goal of sustainable building and construction.

GABS, which was established by the Royal Institute of Chartered Surveyors Foundation with member organisations representing professionals, business, governmental and non-governmental sectors, was formed to accelerate the achievement of sustainable development in the land, property, construction and development sectors and to provide a platform for practitioners in those sectors to contribute to the WSSD. Representatives from more than 40 organisations signed the charter at the summit.

The GABS charter for action

1.0 The global alliance for building sustainability (GABS) is committed actively to promote the adoption of policies and practices to accelerate the achievement of the goal of sustainable development in the sectors of land, property, construction, facilities management, infrastructure and development.
2.0 We affirm the sustainable development vision of the WSSD and seek GABS' inclusion in the subsequent process.
3.0 We are committed to closing the gap between policy and practice and taking practical and determined steps towards making sustainable development a reality for practitioners working in business, government, and/or communities.
4.0 We are committed to creating the opportunity throughout the above sectors in which practitioners are enabled to implement processes and practices that deliver sustainable development.

---

[50] Global Alliance for Building Sustainability (GABS) Charter – http://webapps01.un.org/dsd/partnerships/public/partnerships/51.html

5.0 We undertake to accelerate the achievement of sustainable development throughout the above sectors by actively:

❏ building and strengthening partnerships between policy makers and practitioners;
❏ fostering co-operation and collaboration within and across professions and other stakeholders;
❏ promoting awareness, participation and learning amongst the many stakeholders involved in these sectors;
❏ promoting, supporting and disseminating appropriate research, education and training;
❏ providing a network for collaboration for the member organisations of GABS;
❏ providing a platform for the exchange and dissemination of best and emerging practice;
❏ facilitating the development of tools and performance benchmarks.

6.0 Specifically, the GABS signatories commit to:

❏ promoting the GABS vision, aims and objectives within our individual organisations;
❏ distributing information appropriate to our members about partnerships, processes and practices for sustainable development;
❏ reporting to GABS our progress towards the implementation of the principles of sustainable development.

7.0 In operational terms, we commit to:

❏ contributing to the production of a strategy and business plan for GABS;
❏ supporting the development of GABS.

Finally, we commit to the global alliance for building sustainability and recognise the benefit of partnerships for action, in turn contributing to sustainable development.

Signed for and on behalf of the attached listed organisations at the Indaba Hotel, Johannesburg, South Africa, at the inaugural event for the global alliance for building sustainability, Friday 30 August 2002, within the world summit for sustainable development.

1.3.3   Requirements of facilities managers

Facilities managers play a key role in the development and management of systems and data, primarily through the gathering, analysis and reporting of environmental and social information involving liaison with the rest of the business. The level of this information will be dependent on the importance of CR management within each individual organisation. Some less advanced companies are still collecting

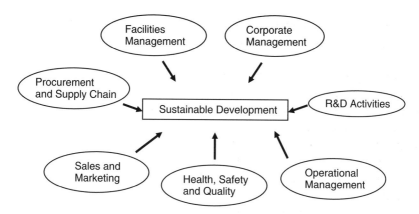

Figure 1.17 The 'ideal' position for business functions to support sustainable development.

what might be described as primary data and auditing their processes against basic legislative requirements. Others who have established effective management and reporting procedures focus more on collecting data to monitor compliance with company policies and objectives.

Within this role, the facilities manager is a key figure to aid and support an organisation in its development and management of a management system (Figure 1.17). In some companies this role has enabled facilities managers to extend their sphere of influence. More and more businesses are moving towards an integrated approach to the management of environmental issues encompassing all areas within the business with the growing appreciation that the environment requires strategic management.

The major focus for facilities managers is to provide added value as part of the management of their property through pinpointing environmental costs and business opportunities. Much of this is through identifying projects such as utility and waste minimisation activities and arguing the case for capital investments which will permit a step change in environmental performance while markedly improving efficiency.

With the additional activities and workload, the facilities manager requires support in order to fulfil the day-to-day and operational management tasks. Most of the activities will rely on the evaluation of incoming utility and waste data, management of contractors and service providers, and the inclusion of environmental criteria within construction and refurbishment projects. The role will also include the management of a team to provide monthly reporting of day-to-day performance, e.g. utilities and waste and incident and non-conformance reporting where poor practices are identified.

The key skills to provide this level of information and detail are communication and management control. These will help to ensure that the team as a whole are aware of the reasons why they are collecting the information and reviewing it on a regular basis. The emphasis will always be on reducing existing costs and understanding changes in consumption, usage or disposals to effect a change at the core of the situation.

The rise in environmental prominence has also brought with it a dearth of publications, magazines and websites to help support and provide guidance to manage environmental issues within the workplace. This spreads across the entire spectrum of environmental management from manufacturing operational activities to multisite property management. In addition, there are now a number of environmental training courses available to walk a manager through compliance, best practice and operational activities.

Properly thought through, sustainable development can be a major market differentiator. The facilities manager will have a major part to play in tailoring and providing service delivery and management processes specifically to reflect an organisation's policy requirements, thus ensuring that these are fully integrated into all aspects of on-site operational activities.

Stakeholders and customers are also having an increasing influence on organisations by requesting them to move towards managing their effects on the environment and society in a more effective and public manner. As part of this, organisations now find that they need to have an environmental policy in place as a minimum, with a mechanism present to implement, measure and monitor the various elements of the policy.

The main environmental risks that a facilities manager needs to manage are described briefly below. This is by no means an exhaustive list, but will provide a description of the varied requirements and responsibilities to be complied with:

❏ Managing contractors and suppliers to ensure environmental legislation and site practices are understood by contractors and suppliers. This will include training and communication at the initial stage, with measuring and monitoring of activities to provide management of the key aims;

❏ Service provision, e.g. cleaning, catering, mechanical and electrical – the inclusion of environmental issues for each service provision including management of plant, use of chemicals, building management system controls;

❏ Construction and refurbishment projects, including the choice of materials, disposal of waste, design of project to carry forward ideas through the various stages of design and build into day-to-day management;

❏ Major capital spend projects, such as chiller unit replacement which should incorporate environmental criteria as part of a life-cycle analysis;

❏ Waste and utility management, particularly the control and reduction of consumption, purchase of green energy, recycling as well as the day-to-day management and legal compliance checks;

❏ Procurement and supply chain management for the service providers and various provision of consumables and specialist equipment to ensure suppliers meet an organisation's environmental criteria; and

❏ Leasing and purchase of property to meet insurance and legislative requirements with due diligence e.g. contaminated land.

The various environmental impacts affecting the services provided by a facilities department are described below in Table 1.3. This is not an exhaustive list but is meant to show the types of activities and the environmental impacts that will affect each service area.

1.3.4   Strategic challenges

Currently there is a small percentage of buildings replaced each year. It is expected that by 2050, a vast majority of buildings standing will be those operating today. If we are to achieve the targets and expectations, the existing portfolio holds a significant role to play. There is a recognition that unless we significantly influence our existing portfolio we will fail to achieve these targets. However, the FM activity has only been recently recognised as an entity in its own right and still suffers from being the poor relative of the construction industry.

Yet FM has little or no involvement in the decisions which result in these impacts, largely the result of front-end capital decisions on the premises. Buildings are commonly handed over with insufficient information, poorly commissioned, or incomplete operation and maintenance manuals due to a variety of factors. FM needs to get more involved at the front end of projects – whether an acquisition, purchase or new build – in order to better manage the resulting environmental and social impacts.

There are a number of strategic areas to better understand the value of FM in the building life cycle and to include FM as part of the decision-making process.

Knowledge management

There is a known gap between the design intent of a facility and its actual operation throughout its life cycle. Part of this is due to changes made during the construction process, part due to poor management of the facility and part due to the lack of knowledge transferred to the FM team on how to effectively manage the facility under the varying conditions. There is a need to better manage the knowledge and change management process from design through to operation to enable properties to deliver their real value. Much of this will be focused upon FM being involved at these initial stages by providing input.

The return circle will enable FM to use their knowledge of the performance of buildings to deliver improved facilities from the design and construction phase, helping to improve the overall asset value and performance of the building stock.

Performance indicators

The definitions of sustainability affecting facilities management are varied and complex depending upon the activity performed, which may in part explain the poor uptake of sustainability by organisations. The development of performance measures will help to support the education process, provide a level of consistency in the marketplace and enable organisations to measure anticipated and actual results. The performance measures will include a level of benchmarking

Table 1.3   Typical measures and issues affecting the FM community mapped against the 14 areas of sustainability

Sustainability category	Typical measures affecting organisations
Management	Achievement of a certified management system, e.g. ISO14001; Formal certification to recognised schemes e.g. LEED or BREEAM; Commissioning of the facility; Consider environmental and social impacts when making decisions about internal financial management systems; Use management systems, which enable the business unit to manage and improve its social and environmental impacts; Use indicators, targets, policies and management processes, which integrate business practices with sustainable development. Monitor and evaluate their success; Review the role and challenge the record of the business unit's relevant senior decision-making body in relation to sustainable development; Publish information about the environmental and social impacts of the business unit.
Emissions to air	Minimise sources of air pollution emissions in the design; Separate sources of pollution from sensitive receptors; Use of low emission finishes, construction materials, carpets and furnishings; Reduce emissions to air, water and land of greenhouse gases, ozone-depleting substances.
Land contamination	Land contamination either recent or historical; Maximise fiscal incentives to remediate; Determine extent of remediation necessary.
Workforce occupants	Reduce nuisance, e.g. noise, odour, visual impact; Promote access to work and services for people with special needs; Encourage people to be involved in giving their ideas and making decisions; Promote employee development; Increase people's opportunities to learn critical skills, e.g. IT, and to keep on learning throughout their work life and personal life; Management of dust, noise during demolition and construction phases of the project.

Continued

Table 1.3 *Continued*

Sustainability category	Typical measures affecting organisations
Local environment and community	Contribute to the local environment and to the local infrastructure to improve safety and provide opportunities for people to become healthier; Working in partnership with local communities; Promote and support community activity and volunteering; Take part in activities which reduce prejudice and promote understanding; Share resources, e.g. equipment, knowledge, skills, and buildings, with community, voluntary, educational or charitable groups; Take part in schemes which utilise the potential contribution to society of people who might otherwise not be employed, e.g. the long-term unemployed, disabled people, and which help all individuals contribute to society and realise their potential; Contribute to the prosperity of the wider local, regional or national economy.
Life cycle or building/products	Inclusion of environmental issues throughout the life cycle of a project to include on-going issues once designed; Use of materials.
Energy management	Management of plant and BMS to affect energy consumption; Generate or use renewable energy; Management of chiller units and air-conditioning equipment, particularly the charge used and following disposal; Control of lighting at night; Review of service charge payment to influence reduction in utility consumption.
Emissions to water	Review potential for sustainable drainage system, e.g. porous pavements swales and basins, ponds and wetlands, infiltration trenches or green roofs; Minimise water use through lean design and operation; Determine potential sources of water pollution and whether suitable processes are in place to minimise the risk of water pollution; Provide flood defences and methods for their protection.
Use of resources	Use materials (including water) and energy efficiency; Reduce use of fossil fuels, metals, aggregates and minerals; Increase use of renewable and recycled materials; Storage of water, its supply and disposal; Storage of chemicals and fuel oils.

Continued

Table 1.3   *Continued*

Sustainability category	Typical measures affecting organisations
Waste management	Reduce waste of all kinds; Legal compliance with the WEEE directive, waste disposal; Collection and disposal of office wastes to increase recycling; Disposal of clinical waste and feminine hygiene; Legal compliance with solid and liquid wastes; Disposal of hazardous wastes such as batteries for an uninterrupted power supply, oils and coolants; Disposal of over-ordered food and consumables; Management of packaging wastes and consumables to reduce at the initial stage and recycle at the end; Management of wastes produced during the project.
Marketplace	Use purchasing power to support the local economy and to support organisations which are contributing to sustainable development; Increase the flow of information to stakeholders; Increase consultation with stakeholders and use it to inform decision-making; Do business ethically and actively opposing corruption and unfair practices; Procurement of consumables including paper, toner cartridges; Procurement of IT equipment for energy consumption and end-of-life re-use; Procurement of cleaning chemicals, pesticides and their effect on the environment; Procurement of locally sourced/ethically farmed foods and purchase of organic products.
Human rights	Promote good health and safety at work, in (global) supply chains and more widely; Promote skills and activities which help people live more healthily.
Biodiversity	Protect and enhance native species and their habitats; Increase the proportion of resources that are from sustainable-managed sources, e.g. not over-fished, over-harvested, clear-felled or taken from the wild.
Transport	Use of video and teleconferencing for meetings; Environment-friendly choice of vehicles for company cars, with a limit for CO_2 emissions; Reduce distance travelled by suppliers and contractors to provide services.

and correlate the performance of the building as intended in design stage through to operation. Current tools exist which measure the sustainability performance prior to occupation, but these measures are not continued into operation.

Whole-life value

There is a barrier between the capital and revenue budgets which exists and under-values the delivery of exceptional buildings from a sustainability perspective. By its very nature, whole-life value will look to define and measure tangible and intangible costs and benefits for a project, within reasonable boundaries, to enable option appraisal and comparisons to be made. There is a need for greater consistency in how value is measured and formal guidance on the structure and process used.

There is a strong focus by organisations on cost reduction to support share price, with facilities seen as a primary area of overspend and therefore a cost burden to be reduced. This leads to challenges in the delivery of a long-term programme for a facility in the light of reducing capital and revenue spend. This is highlighted in the split between construction and operation costs, which are budgeted differently, commonly managed separately, and at odds with each other. Control, underspend and delivery within timeframes are critical measures for both which do not promote co-operation. However, life-cycle benefits are commonly missed through such a narrow view of the building programme.

There is a need to change the way wealth is measured and these budgets are delivered if the vision of sustainable buildings as the norm is to be realised. Rather than just a financial metric, the environmental and social impacts and benefits to a scheme need to be incorporated as a cost to provide a full financial profit and loss account.

Skills and competency

It is recognised that there is a lack of skills within the FM industry to actively manage facilities from a sustainability perspective. In part, this is due to wide experiences from which the facilities management community enters the profession and due to the wide range of skills required. In order to meet the vision, a greater involvement of FM within the design and construction activities will require the skills to communicate ideas effectively, and to understand the process in which these individuals are operating within.

Increasing liability

Organisations and increasingly those managing properties are holding greater responsibility for the actual and perceived liabilities of the activities performed. Partly governed through legislation and by public perception, liability covers the environmental, social and economic aspects of a business in the local communities where the facility is based and the market within which it operates.

FM has a great responsibility for compliance with many of the legislative requirements noted due to the management and control over issues such as waste disposal, discharge of effluent and monitoring and measurement of plant equipment. Changes to legislation will mean that individuals are liable for environmental damage, similar to that for health and safety incidents. In this way not only have managing directors been sentenced to imprisonment, but those lower down the line have been found in breach of environmental legislation.

'Contrary to common perceptions, higher environmental standards in industrial countries have not tended to lower their international competitiveness.'
World Bank paper 'Competitiveness and Environmental Standards'[51]

'We found no strong evidence that environmental regulation destroys jobs and businesses ... with compliance costs averaging 1–2% of business turnover, such costs are unlikely themselves to shift competitive advantage significantly ... [environmental factors] do not suggest that promoting lower standards at home would be to the competitive advantage of British business.'
Confederation of British Industry[52]

These statements are in direct conflict with the messages put out by various industries and professional bodies promoting business. Groups have complained that meeting reductions in carbon dioxide and other environmental contaminants would cost jobs and make industry uncompetitive. This is the primary reason for the United States not signing the Kyoto Protocol. It is not uncommon for the messages spread by business on the implications of new legislation to be dramatised to extenuate the potentially devastating impacts to industry:

❑ The EU car industry predicted that catalytic converters would cost £400 to £600 per vehicle. The real cost was £50 with over £2 billion of societal benefit in the United Kingdom alone;
❑ The United Kingdom's electricity industry predicted that EU laws to combat acid rain would increase the cost of electricity by 25%. The actual cost was negligible;
❑ The US automotive industry decided to postpone the development of dual fuel (petrol and electric) vehicles due to perceived low demand. In 2006, Toyota will double production of the Prius and will become the world's largest automotive provider by volume.

[51] World Bank paper "Competitiveness and Environmental Standards – http://www.worldbank.org/
[52] Confederation of British Industry – http://www.cbi.org.uk

2 Sustainable business management

Sustainability as part of facilities management is as much delivered through the management systems and processes as it is by the programmes and initiatives. To become a way of life, built into day-to-day activities, there is a need to understand not only the FM strategy, but also the business strategy for the organisation.

This chapter will cover:

Green service management
FM has a critical role in the supply chain to extend corporate responsibility (CR), the business drivers behind this, and the methodology to incorporate sustainability elements into day-to-day activities:

❑ Introduction to green supply chain management, including the drivers and benefits;
❑ Describe the business reasons for implementing green/ethical supply management;
❑ Methodology to implement for new and existing suppliers and contractors; and
❑ Templates and proforma.

Environmental legislation
An overview of the legislative processes and requirements that apply through global, regional and local regulatory requirements. The section also provides the general position to be applied to minimise risk and manage corporate responsibility impacts:

❑ Overview of legislation affecting the European Union, United Kingdom, United States and Asia Pacific;
❑ Risk management and control affecting air, land and water; and
❑ Role of facilities management in delivering legislative compliance.

Business case and corporate challenges
The drivers for CR provide the basis from a corporate perspective to enable the facilities manager to develop an effective business case to deliver CR aligned and supportive of business needs:

❑ Drivers for business and organisations from a global and local perspective;
❑ The main benefits and drawbacks to the implementation at a business level;

❑ How to develop an effective business case and management programme; and
❑ Implementing the business case and delivering improvement.

Corporate responsibility management systems
This section reviews the growth in implementing environmental and social management systems, the business benefits and reasoning behind this, as well as providing an overview of implementing a corporate responsibility management system (CRMS):

❑ Introduction to management systems, how they operate and how to get the best from them;
❑ Definitions and distinctions between environmental, social and corporate responsibility management systems;
❑ Advantages and disadvantages of management systems; and
❑ Implementation of a social and environmental management system.

Corporate responsibility reporting
This section reviews the development of corporate responsibility reporting and the role FM can play in delivering these reports both internally and publicly:

❑ Background to CR reporting including the drivers;
❑ Implications of the operating and financial review;
❑ Understanding the needs and requirements of stakeholders; and
❑ Developing a corporate responsibility report.

2.1   Green service management

The role of procurement and managing suppliers and contractors is one that is generally overlooked when assessing environmental and social risks, with only economic issues generally taken into account. There are many reasons for this, including the difficulty in assessing and quantifying the impact procurement has, and the complexities involved in providing standardisation across the various types of contracts in place. It is important to note that management of the service as described in this section is equally applicable to an in-house service provision.

Organisations will procure everything, from buildings through leases, major contractors such as M&E (mechanical and electrical) through medium- and long-term contracts, and smaller contractors or suppliers for one-off projects or relatively low-level activities. To try and incorporate environmental and social requirements into this wide ranging remit is not enviable. However, the inclusion of these requirements can have great benefits including reduced costs, improved employment conditions, verification of best practice through the supply chain and greater corporate visibility.

Section learning guide

This section reviews the role of the supply chain to extend CR, the business drivers behind this, and the methodology to be able to incorporate sustainability elements into day-to-day activities:

❑ Introduction to green supply chain management, including the drivers and benefits;
❑ Describe the business reasons for implementing green/ethical supply management;
❑ Methodology to implement for new and existing suppliers and contractors; and
❑ Templates and proforma.

Key messages include:

❑ Start with the most important suppliers – cost, risk, importance, impact;
❑ Use information to encourage change – can influence the performance of suppliers and contractors;
❑ Local contractors and suppliers should be considered;
❑ Identify what information is required and how it will be used – collecting information for no use is wasteful for yourself and your suppliers;
❑ Explain to suppliers why how they operate is important and provide feedback; and
❑ Develop a long-term relationship – well managed gradual change will provide better returns than a fast and careless approach.

2.1.1   Introduction

The management and operation of buildings has increasingly become outsourced and out-tasked to a series of contractors operating on site. Ranging from single individuals through to multi-national organisations, the provision of cleaning, catering, security and other non-core activities are taking place throughout the day.

One of the largest risks many companies face, especially those who have 'outsourced' their service activities, is how assurances can be obtained from the supply chain to meet an organisation's brand values. As Nike and others have proved, your brand is just as much at risk by the actions of your supply chain (Section 1.2.5). The degree of risk will vary greatly depending on the nature of the goods or services being provided. In general, the more complex the goods or services together with the degree of direction the supplier is taking from the client, the greater the level of risk. In general, roles with little communication and involvement with the client will have difficulties.

The FM industry employs some of the lowest paid staff through cleaners, security guards and caterers within an organisation working unusual hours either on a permanent or contract basis. The reasons for working are more often to make ends meet from a secondary wage. The role of FM is intrinsically linked to supplier management and control where most of the service activities are performed and sustainability damage takes place.

LIVERPOOL JOHN MOORES UNIVERSITY
LEARNING SERVICES

In addition, the provision of goods and materials has become a global enterprise through the movement of suppliers to lower-cost countries in eastern Europe and South-East Asia. Everything from clothing, electronic items and furniture are manufactured abroad due to cheaper labour and assembly costs which reduces overheads and sales costs.

An organisation's supply chain can influence and impact upon a number of areas affecting the social aspects of the individuals working on site and throughout the supply chain:

❑ Wages paid to suppliers and contractors' employees;
❑ Working hours and conditions of employment;
❑ Human rights;
❑ Contractor and supplier management;
❑ Payment to terms and conditions, in particular payment on time; and
❑ Benchmarking of supplier performance.

There has been an increasing focus on working conditions and controls on suppliers on a global basis through monitoring and publications from non-governmental organisation's such as Oxfam and Save the Children, corporate reporting programmes and the development of procurement policies.

2.1.2 Best value

Historically, the FM procurement process has been focused upon the lowest cost service provision at a point in time, without regard for later costs and caveats incurred. The outsourcing and out-tasking of services was primarily seen as a means to reduce headcount and to save costs. The construction industry has strongly adopted this approach, with contracts provided on less than 2% margin with additional requirements making the profit up to 6%. Part of this difficulty has been due to the relatively short contract timescales of less than three years which limits innovation with paybacks over this timeframe.

Purchasing needs to move away from 'cost is king' thinking when looking at tenders and take a more holistic, long-term approach, basing decisions on true value rather than the price tag alone. One of the tasks is to work out how to add apples to pears to oranges – environmental considerations to social considerations to economic considerations.

The UK government has implemented a strategy to incorporate sustainable purchasing requirements within its £125 billion spend each year with the target of becoming a world leader in sustainability by 2009. The intention is to stimulate the local market within the United Kingdom and Europe to produce increasingly sustainable materials at cheaper prices to compete directly with current products. Benefits are anticipated to include not only more sustainable procurement but also a rise in the quality of materials and products provided.[1]

---

[1] Sustainable Procurement Task Force – http://www.sustainable-development.gov.uk/government/task-forces/procurement/index.htm

To achieve this, the UK government is working with the supply chain through a management programme to develop and implement sustainable solutions for its product and services. In particular, there is a need to incorporate sustainability principles within the construction and refurbishment of all buildings. Such an approach should be evaluated against the life-cycle costs rather than the initial capital cost – therefore the product or service which provides the best overall value, rather than an upfront saving but longer operational costs, would be chosen. Where possible, the environmental and social benefits should be included in addition to the direct financial metrics.

To further support the initiative the UK Office of Government Commerce (OGC) has published on its website a list of 'quick win' products and services that have been shown to meet high environmental standards and give value for money. The list will be kept under review and its coverage widened over time.[2]

2.1.3 Benefits and issues

The business case for going the extra step and reviewing the life cycle impacts of the materials, products and services purchased is well documented.[3] The role of working in partnership with suppliers and contractors to reduce the environmental and social burden can have significant benefits. There is a strong business case for integrating environmental and social considerations into all stages of the purchasing process in order to reduce the impact on human health and the environment to support an organisation's corporate governance.

Reputation: A company's commitment to social and environmental responsibility influences consumer patterns and the decision to do business with a company. When buying a product or service, 70% of European consumers say this is important, and one in five would be very willing to pay more for products that are socially and environmentally responsible.[4] Analysts have calculated that the Co-operative Bank's ethical and ecological positioning has made a sizeable direct contribution to its profitability of around 20%.[5]

Brand equity: 85% of firms consider brands to be their most important asset – they can be shown to be a source of sustainable competitive advantage and value creation.[6] It is estimated, for example, that 96% of Coca Cola's value are intangibles – reputation, knowledge and brand, Kellogg's is 97% and American Express 84%.[7]

---

[2] UK Government Procurement – http://www.ogcbuyingsolutions.gov.uk

[3] Tukker, A., Eder, P. and Suh, S. 'Environmental Impacts of Products', *Journal of Industrial Ecology*, MIT Press, 2003; Lippke, B. *et al.* CORRIM: Life-Cycle Environmental Performance of Renewable Building Materials, Forest Products Journal, 2004 – http://forestprod.org/

[4] *The first ever European survey of consumers' attitudes towards Corporate Social Responsibility + country profiles*, 2000, CR Europe 2000 – http://www.csreurope.org

[5] Co-Operative Bank – http://www.co-operativebank.co.uk

[6] Templeton Briefing 5: Brand Risk Management in a Value Context, Marsh and McLennan Companies, June 2000 – http://www.templeton.ox.ac.uk/templetonviews/docs/2000-03.pdf

[7] Interbrand Best Global Brands 2000 – http://www.interbrand.com

Share value: Socially and environmentally responsible procurement makes for a better investment prospect. The number of socially responsible investment (SRI) funds is growing rapidly, accounting for nearly 13% of the $16.3 trillion in investment assets under professional management in the United States. Over £80 billion of UK equity held by UK occupational pension funds are subject to engagement undertaken by UK fund managers in line with the pension funds' policies covering environmental and social issues. Institutional investors are prepared to pay a premium of more than 20% for shares of companies that demonstrate good corporate governance.[8]

Innovation: Sustainable supply chain management can stimulate investment in new sustainable technologies, help develop less harmful products and establish markets for new, more sustainable products and services. Rising operational costs resulting from environmental taxation can be counteracted by good supply chain management. The EU waste electrical and electronics directive will force more creative ways for disposal of spent electrical goods.

2.1.4 Developing a procurement policy

The development of a procurement policy should be aligned with the CR programme directly and include environmental and social criteria. It should define the overall strategy and objectives, and convey a strong clear message to suppliers and contractors about what the organisation expects from them. These requirements will change and be different for each organisation depending upon the nature of the business, supply chain relationships and size.

The most effective procurement policy will operate in the same way as that of the environmental or corporate responsibility policy – through senior management support and a robust management system underpinning and delivering the essence of the policy statement. Typical areas to include are:

❏ A commitment to environmental and social inclusion within the supply chain management and purchasing strategy;
❏ A programme to initially benchmark and set parameters to achieve continual improvement and progress;
❏ The programme is to be in keeping with the operations of the business, items procured and main areas where greatest benefit can be provided;
❏ Communication and development with the suppliers to achieve the programme;
❏ Ability to measure progress and achieve the targets coupled with the provision of reporting; and
❏ Auditing of the environmental and social supply chain management process.

This will mean the involvement of the procurement professional and importantly integration with the processes and systems to buy into the benefits and wider corporate objectives and targets.

---

[8] McKinsey & Company – http://www.mckinsey.com/

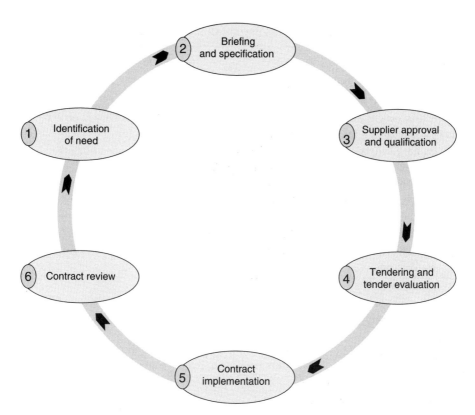

Figure 2.1   Procurement cycle (adapted from Barbara Morton, UMIST).

2.1.5   Procurement of new suppliers and contractors

The typical process for procurement activities is through a cycle, as shown in Figure 2.1, describing the evolution of understanding the need, developing a specification, tendering the activity, implementing a contract and measuring that the contractor ensures effective control of the service. The key is to incorporate the operation into day-to-day activities. The documentation which is commonly used in conjunction with these various steps is shown in Table 2.1.

Identification of need

The need for a new service identified within FM can relate to most activities and services – either as a new innovation or the re-tendering of a contract. When a need has been identified it is important at this stage that environmental criteria are incorporated into the overall objectives of the service, particularly for larger service delivery – see Table 2.2.

Based on the CR policy or a specific procurement policy, the key objectives can be identified to review the potential impacts from the delivery of the service. Examples are provided in Table 2.2. Against the sustainability categories, typical impacts are provided as outcome based requirements. Through focusing on these impacts greater control can be obtained and the sustainability risk minimised.

Table 2.1   Documentation used during the procurement cycle

Steps represented	Related documentation
(1) Identify a need of the company for a specific service.	Policy objectives
(2) A brief and specification are developed for the service that is required.	Brief and selection criteria
(3) Potential suppliers are identified and some formal or informal qualification of suppliers follows before best quotes are obtained.	Tender specification
(4) The best quote is identified through a tendering process.	Contracts
(5) The purchase is made and the service delivered.	Reporting
(6) Finally, some form of contract review is conducted.	Performance evaluation

Briefing / specification

At the end of the first stage, a comprehensive list of the environmental impacts of the service and the policy ideals will have been developed. The next stage is to develop a briefing and details for any pre-tender specification or request for information (RFI). This information should be incorporated into existing documentation where possible. This will provide the opportunity to avoid rather than minimise the environmental impacts. It is difficult to define how environmentally friendly a product is, but eco-labels such as the FSC (Forest Stewardship Council) and Energy Star should be used with care as these labels only provide an indication of how sustainable a product is.

The briefing stage should include the encouragement of suppliers to be proactive in the aims and objectives identified for the service as a vision for how the organisation wishes it to be performed. This will include all elements of service delivery including quality, customer care and environment. It is therefore important that the sustainability impacts identified earlier form part of the pre-tender process so that the service provider can gauge the level of work required at the outset.

In addition to this vision of the service, a series of questions should be developed to identify the nature of environmental management for any provider. A typical list of questions to cover corporate responsibility within a RFI is shown below:

❑ Has your company published a clear definitive environmental/sustainability/ social/corporate responsibility policy statement? (Delete as applicable)
❑ Is this policy statement supported by a comprehensive and detailed management system which meets the requirements of ISO14001/EMAS/AA1000/ SA8000?
❑ Does the policy/management system have the full support of the company board and has it been endorsed by the CEO or MD?

Table 2.2 Impacts detailed for the identification of sustainability objectives

Sustainability category	Potential impact
Emissions to air	CO_2 from building energy use; Other emissions from the building
Emissions to water	Water pollution
Use of resources	Procurement of materials; Use of materials from local sources; Embodied energy from materials chosen
Energy management	Electricity use; Energy use/management; Heating fuel use; New and renewable energy
Waste management	Re-use of materials directly or as secondary products; Recycle materials to avoid landfill; Waste management technology; Responsible and effective disposal of waste generated
Land contamination	Contaminated land management; Oil/fuel leaks
Local environment and community	Community engagement; Community health and wellbeing; Nuisance
Workforce occupants	Recruitment and retention of staff; Training and development needs – skills and knowledge; Management of staff; Health and safety
Life cycle of products and services	Material efficiencies over lifetime – design and lean construction; Extend durability of existing structures; Life cycle cost analysis and climate change implications; Construction and operational technology
Marketplace	Cost effectiveness/profitability and construction time/budget; Benchmarking of progress; Payment of suppliers; Supply chain sustainability practices and performance
Human rights	Flexible working; Health and safety; Stakeholder consultation
Biodiversity	Value of habitats, species and land; Safe areas for protected species and habitats; Access to the local community; Construction impact on habitats and species
Transport	CO_2 from transport to/from building; Other emissions from transport

❑ Have defined roles and responsibilities for competent individuals and committees been established?

❑ Does the company provide and review employee training to develop competency in each appropriate job related activity?

❑ How do you think you compare against your peer group?

❑ What sustainability criteria do you measure/benchmark and how effective has this been to improve performance?

❑ Can you provide examples of where you have supported delivering a client's sustainability issues?

❑ How do you engage with your supply chain on sustainability – provide examples of where this has been effective?

❑ What training and education do you provide to staff on sustainability? Is this for the supply chain as well?

❑ Who will be responsible for sustainability on this account both on-site and off-site?

An alternative to a series of questions is to request that suppliers document their achievement against a roadmap capturing the main areas of corporate responsibility. An example is provided (Figure 2.2) highlighting ten categories for which a scoring ranging from one to five has been devised – where one represents poor practice and minimal implementation and five represents best practice. This self-scoring system operates well, with the ability to audit achievement against the roadmap prior to contract, and provides the ability to request annual improvement against the roadmap.

Tendering

Dependent on the type of service required, a pre-tender process may not be required, or undertaken. A full tender process or request for proposal (RFP) to a service provider will provide the individual elements of a service to enable accurate costs and provisions to be provided. To incorporate environmental requirements at this stage is critical, and these requirements must be maintained throughout the contractual negotiation phase.

Generic statement Within the tender, information should be provided to describe the importance of corporate responsibility within the organisation. This can be in the form of a generic statement in the opening sections and therefore be standardised for all tenders. An example is provided below from Save the Children's specifications which links together the aspirations of the building and the values of the charity.[9]

Save the Children works to help build a better world for present and future generations. We are aware of the impact our buildings have on the environment and on the people who use and manage

---

[9] 2004 tender documents for Save the Children Fund HQ building, not published.

	1	2	3	4	5
Management decisions	Sustainability considerations are not included at management level.	Partial roles defined, though mainly ad hoc approach to considerations.	Management consider key issues through reviews.	Structure, responsibilities and environmental budgets limited to information provided to management.	Proactive role by management in reviewing the financial, emergency and structure of the business.
Legal compliance	No means to monitor or provide evidence of compliance with legislation.	Reactive monitoring and identification of requirements.	Identification and response to major areas of legal concern.	Documented compliance to legislation, and communication to internal and external bodies.	Proactive response to legal compliance, regulatory bodies and the local community.
Life-cycle processes	No environmental issues evaluated.	Key areas undergo an ad hoc review.	Major items are reviewed and some communication with suppliers to meet legal requirements.	All items or major changes are identified and evaluated prior to purchase. Storage, training and use is also considered.	Life-cycle assessments made for all items, with the inclusion of raw material usage.
Continuous improvement	Reactive performance to rectify incidents. Policy statement not able to define future progress.	Progress generally defined within the policy. Monitoring to avoid incidents is ad hoc.	Major areas are monitored, assessed and reviewed. Some feedback to remediate incidents.	General progress documented and defined. Monitoring, assessment and reviews communicated to management.	Systematic and documented process to provide a proactive response to identifying and resolving issues.
Supply chain management	No considerations taken during selection, monitoring or control.	Ad hoc process for reviewing supply chain throughout each phase.	An adequate process to review suppliers for most issues.	A documented process to review suppliers throughout each phase.	Robust process to work with supply to improve the overall impacts.
Communication and education	Minimal communication takes place with little involvement by parties.	Limited training and involvement takes place of interested parties. Only main issues are reported.	Communication and education processes in place though information is relatively limited.	Internal and external aspects have been considered for most aspects.	Procedures and plans to provide comprehensive communication and education to both internal and external sources.
Marketplace	Little knowledge of external standing with a view to sustainability.	Partial or reactive reviews of performance against others.	Structured reviews made against non-specific or generic groups.	Tailored reviews against critical areas to focus on performance and feedback.	Proactive transparent assessment with stakeholders against all impacts.
Waste and resources minimisation	No ability to identify waste arisings or generation.	Major or ad hoc data collection to focus on key arisings.	Strong data collection to focus on end-of-pipe solutions.	Processes in place to minimise waste arisings throughout material life cycle.	Systematic approach to minimise material use and consumption on nature and impact.
Utilities and Infrastructure Management	No ability to identify energy consumption or costs.	Data collection of costs and consumption with limited activity.	Ad hoc review of savings, with implementation of no cost options.	On-going collection of consumption and costs to highlight inefficiencies and savings.	Systematic cost and consumption review against building usage.
Biodiversity and Land Management	No assessment made of land impacts.	Ad hoc response to issues affecting land.	Review of major impacts and value with short-term solutions implemented.	Process to review all impacts and habitats to maintain and improve.	Structured programmes in place to improve land quality.

Figure 2.2 Corporate responsibility roadmap. (© Jacobs UK Ltd, 2006).

them and want to procure services in a way that contributes to reducing this impact. We expect contractors to contribute to the sustainability aims for our new London office by promoting good working practises and minimising the environmental impacts of their business.

Save the Children would like to develop a partnership approach with contractors providing services in the building. The contractor will interface with others while fulfilling their contract, including the Local Authority, the public, local residents and businesses, the tenants within the building and staff and visitors to the building. They are expected to do this in a way that is compatible with Save the Children's values. Save the Children is concerned about the differential minimum wage level set for workers under the age of 21, and is developing a policy on this.

Specific statement Further to the generic statement, a series of specific criteria should be provided that focus around the environmental and social impacts which the service is intended to have. It is important to list out the requirements and criteria that you want from the service provider to ensure the cost for these activities is built into the bid response. Examples are provided below:

Address CO_2 emissions from building energy use, taking all actions necessary to:

❑ Gather basic relevant information;
❑ Passively inform and improve occupant behaviour; and
❑ Implement no/low cost measures to reduce impacts.

Improve the quantity of waste recycled, taking all actions necessary to:

❑ Monitor, benchmark and develop targets;
❑ Actively educate occupants and improve behaviour; and
❑ Implement measures with a simple payback of <2 years to reduce impacts.

Methods of introducing environmental criteria into calls for tender are explained in clear, non-technical language in EU documentation.[10] Particular emphasis is laid on the need to take into account the cost of the life cycle of purchased products, and the manual's publication follows the implementation of directives that aim to encourage public authorities to introduce ecological criteria in their tendering processes.

An example is provided in a case study of Oxfam's ethical purchasing policy and supplier questionnaire. Oxfam have spent many years developing and revising the formats of both these documents and encourage suppliers to act together to meet the core aims of Oxfam.

Oxfam have permitted the use of these documents and encourage other companies to use the information to improve supplier performance and ethical purchasing. For further information see http://www.oxfam.org.uk/suppliers.

---

[10] Green Public Procurement by the European Commission – http://ec.europa.eu/environment/gpp/index.htm

Case study: Ethical purchasing – Oxfam GB

Much of Oxfam GBs programme of work seeks to achieve sustainable development for people living in poverty. But Oxfam also needs to be aware of the impact of all its activities on the environment and on communities. Every project, operation or activity will, as part of its management routines, consider its impact on the environment and communities. These management routines must be set up to be both cost-effective and sustainable.

Oxfam's activities impact on the environment and communities through its:

❑ Supply chains' employment conditions;
❑ Direct and indirect use of non-renewable carbon fuels in its buildings and for passenger and freight transport via road and air;
❑ Use of scarce/non-renewable raw materials in the supply chains of the items it purchases;
❑ Use of harmful materials or high energy consuming processes in the production of the items it purchases;
❑ Disposal of waste products;
❑ Relationships with local communities where activities take place; and
❑ Welfare of staff, visitors, volunteers, neighbours and other stakeholders.

Oxfam will follow the principles of *Reduce, Reuse, Repair and Recycle* (4Rs) in managing its environmental impact. Using the 4Rs will not only minimise environmental impacts, it also makes good economic sense.

Oxfam recognises that globalisation of trade means that more and more of the goods and services we buy are at risk of being produced by workers in unregulated environments. Our mission is to work with others to provide lasting solutions to poverty and suffering. Promoting internationally recognised human rights, which include labour rights, is at the heart of the work we do.

Oxfam has been at the forefront of campaigning for, and promoting, ethical trade. Oxfam believes that the companies we buy goods and services from should take a similar responsibility. That's why compliance with our code of conduct, within a reasonable length of time, is a consideration in all supplier relationships. At Oxfam we believe we have responsibilities to all of our stakeholders – staff, volunteers, donors, and the people who ultimately benefit from our activities. So we need to ensure that all our resources are used effectively and appropriately – and that includes the money we pay to you. We also recognise that our approach must be realistic and manageable for you as well as for us. Achieving significant results means working in co-operation with you.

What we ask is that you:

❑ Demonstrate commitment at director level to: continuous improvement in your own company, towards compliance with the code of conduct and in the identification of risks within your supply chain;

❑ Work with your suppliers so that they do the same, preferably by adopting a code of conduct;

❑ Report progress annually, e.g. by completing *our supplier questionnaire*; and

❑ In return, you can expect guidance and support in implementing the code.

By adopting an ethical purchasing policy, Oxfam is setting a positive example for the wider business community.

Oxfam ethical purchasing policy

Oxfam's policy is to seek to purchase goods and services which:

(1) are produced and delivered under conditions that do not involve the abuse or exploitation of any persons; and

(2) have the least negative impact on the environment.

Such considerations will form part of the evaluation and selection criteria for all goods and services purchased by Oxfam. In addition, Oxfam will seek alternative sources where the conduct of suppliers demonstrably violates the basic rights of Oxfam's intended beneficiaries, and there is no willingness to address that situation within a reasonable time period, or where companies in the supply chain are involved in the manufacture or sale of arms in ways which are unacceptable to Oxfam.

Purpose of the policy

The purpose of the policy is to:

(1) promote good labour and environmental standards in the supply chains of goods and services to NGOs; and

(2) protect Oxfam's reputation.

Code of conduct for suppliers

Suppliers adopting this code of conduct should commit to continuous improvement towards compliance with the labour and environmental standards specified, both in their own companies and those of their suppliers.

Labour standards

The labour standards in this code are based on the conventions of the International Labour Organisation (ILO):

❑ Employment is freely chosen;

❑ Freedom of association and the right to collective bargaining are respected;

❏ Working conditions are safe and hygienic;
❏ Child labour shall not be used;
❏ Living wages are paid;
❏ Working hours are not excessive;
❏ No discrimination is practised;
❏ Regular employment is provided; and
❏ No harsh or inhumane treatment is allowed.

Environmental standards

Suppliers should, as a minimum, comply with all statutory and other legal requirements relating to the environmental impacts of their business. Detailed performance standards are a matter for suppliers, but should address at least the following:

❏ Waste management;
❏ Packaging and paper;
❏ Conservation; and
❏ Energy use.

Business behaviour

The conduct of the supplier should not violate the basic rights of Oxfam's intended beneficiaries. The supplier should not be engaged in:

(1) the manufacture of arms; or
(2) the sale of arms to governments which systematically violate the human rights of their citizens; or where there is internal armed conflict or major tensions; or where the sale of arms may jeopardise regional peace and security.

Operating principles

The implementation of the code of conduct for suppliers will be a shared responsibility between Oxfam and its suppliers, informed by a number of operating principles, which will be reviewed from time to time.

Qualifications to the policy statement

The humanitarian imperative is paramount. Where speed of deployment is essential in saving lives, Oxfam will purchase necessary goods and services from the most appropriate available source. Oxfam can accept neither uncontrolled cost increases nor drops in quality. It accepts appropriate internal costs but will work with suppliers to achieve required ethical standards as far as possible at no increase in cost or decrease in quality.

Implementation with suppliers

The code is implemented with all suppliers through a three-phase process to assess all new incoming suppliers, work with those chosen to sequentially improve their practices, and audit the high-risk suppliers (Figure 2.3).

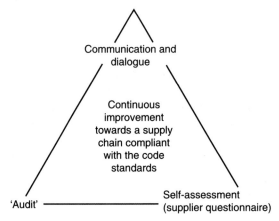

Figure 2.3   Oxfam's responsible supply chain code.

(1) Risk assessment is provided as a supplier questionnaire completed by potential suppliers as a self-assessment. The responses are reviewed and a risk rating provided by trained people. Where a medium or high risk is identified, additional steps are taken.

(2) Dialogue is critical and a strong emphasis is placed at the outset of supplier relationship before selection, e.g. annual supplier days.

(3) Labour and environmental assessment – a number of key suppliers are identified to be audited and categorised upon a risk based criteria: risk of poor standards, visibility, spend, criticality, leverage, SQ (services quality) rating. A formal assessment is performed with notice provided, a structure to the assessment, interviews with workers and feedback provided. In total, only 5–10 are performed each year, mainly in the United Kingdom, though some third-party audits are performed, e.g. in India/China.

Oxfam

For office use only	
Ethical Risk*	
Risk rated by	Date:
EIRIS checked?	
* Purchasers see intranet for process	

OXFAM GB SUPPLIER QUESTIONNAIRE

Thank you for taking the time to complete this questionnaire, we ask all our suppliers to do this. The form should be completed by a senior manager responsible for governance and ethics. To view the Oxfam Ethical Purchasing Policy and Code of Conduct for Suppliers and to download this questionnaire, visit our website at www.oxfam.org.uk/suppliers.

Oxfam staff member you have main contact with: Send the completed form to this person.

Name	Department	Office location

PART A: About your company

1. Name of the company and any parent or subsidiary:

2. Name and position of person responsible for governance/ethics:

3. Name and position of person completing this questionnaire if different from above:

4. Contact Details:

Company address
Telephone number: Fax number:
Email address: Website:

5.   Location of other operational sites and their functions with approx number of employees:

6.   Type of product or service you supply, or might supply, to Oxfam GB:

7.   Company's main customers:

8a.   How many people does the company employ? _____   8b. Approx % of women: _____%

9a.   Do you use home workers in any part of your business?   ☐Yes   ☐ No

9b.   If Yes, approx how many? _____   9c.   What processes do they carry out?

10.   What is the legal status of the company supplying Oxfam?

☐ public limited company   ☐ private company limited by guarantee   ☐ partnership

☐ sole trader   ☐ not-for-profit organisation   ☐ government agency

☐ other

11a.   Company turnover in £ or local currency _____   11b. Turnover of the part of the

business that would supply Oxfam £_____   11c.   Oxfam business as a % of 11b if known.

_____

PART B: Operational Standards

12.   Does the company have any of the following policies/statements?   If Yes, please attach a

copy.

☐ quality   ☐ health & safety   ☐ equal opportunities   ☐ training and development

☐ company ethics   ☐ environmental management   ☐ diversity   ☐ social objectives

13a.   Are employees free to join or form a trade union? ☐Yes   ☐ No

13b.   Do any employees belong to a trade union?   ☐Yes   ☐ No

13c.   Please specify any trade union/s you recognise for the purpose of collective bargaining:

14. Are any other forms of representation used? If so, give details below.

☐ staff association   ☐ works council   ☐ elected Health and Safety Committee

☐ workers co-operative ☐ employee share ownership   ☐ other

| |
| |

15. How do you ensure employees are aware of their rights?

☐ written contracts   ☐ employee handbook   ☐ staff notice boards   ☐ Intranet   ☐ other

16. Please indicate ways in which you consult employees about decisions which affect them, and get

their feedback or ideas

☐ meetings   ☐ surveys   ☐ suggestions box   ☐ other

17a. How much do the lowest paid workers earn in relation to the minimum wage?

| |
| |

17b. Is anyone paid piece rate?   ☐Yes   ☐ No

18. What are the normal weekly working hours for employees?   _____

19a. Is overtime voluntary? ☐Yes   ☐ No   ☐ Sometimes   ☐ Not applicable i.e., no overtime worked

19b. Is it paid at a premium rate? ☐Yes   ☐ No   ☐ Time off in lieu given   ☐ depends on employee

20. What is the youngest age at which someone can be employed by the company? _____

21. Does the company have any recognised Operational Standards for products supplied to Oxfam?

	Quality e.g. ISO9000	Environment e.g. ISO14001	Labour e.g. SA8000	Training e.g. Investors in People
certified to:				
working towards:				
Other standards:				

22. Were any Health & Safety risk assessments carried out last year? ☐Yes ☐ No Give details

_____

23. Has the company had a labour standards audit carried out? ☐Yes ☐ No Give details

_____

24. Is your company committed to achieving the labour and environmental standards in Oxfam's Code of Conduct for Suppliers? ☐Yes ☐ No

PART C: Sourcing from your suppliers

25. Which of the following do you assess suppliers against?

☐ quality ☐ financial ☐ labour ☐ production ☐ management ☐ environment

26a. Do you have a Policy/Code of Conduct for sourcing goods/services? ☐Yes ☐ No

26b. If Yes, does it include standards relating to: ☐ labour? ☐ business ethics? ☐ environment?

Please attach a copy.

27. How do you assess suppliers? ☐ questionnaires ☐ visits ☐ audits Please give details.

_____

28. Do you subcontract/ outsource production of goods or services? ☐Yes ☐ No If yes give details.

_____

29. List the locations (areas/countries) of your key suppliers and give an outline of the supply chain for the products or services relevant to Oxfam, or attach a supply chain map. Include any home workers.

_____

30. Is your company committed to assessing labour and environmental risks in your supply chain?

☐Yes   ☐ No

PART D: Continuous improvement

31. What progress has your company made this year relevant to the standards in Oxfam's Code of Conduct, and what will you do you next year?

a) In your company	Progress
	Plan
b) In your sourcing from suppliers	Progress
	Plan

Name:	Signature:	Date completed:

NB There are some industries Oxfam has run campaigns on to highlight the harm they can cause to poor communities. If your company, or any parent or subsidiary, has any involvement with production of sales of weapons, infant formula or pesticides, please tell your Oxfam contact.

Examples of continuous improvement plans suppliers have sent which are acceptable to Oxfam:

a) In your company	Make a written statement of intent to show our Company's policy and publish our code of conduct
b) In your sourcing from suppliers	Write to our suppliers and explain the value of this process and encourage them to respond with a statement of intent

LIVERPOOL JOHN MOORES UNIVERSITY
LEARNING SERVICES

a) In your company	
b) In your sourcing from suppliers	Produce a Supplier Questionnaire to include labour standards
a) In your company	We think we comply with your code of conduct
b) In your sourcing from suppliers	Each supplier will be made aware of your code of conduct
a) In your company	We are a small company formed out of a larger one. We do adhere to an unwritten code of ethics by buying Fair Trade supplies and recycling paper, print cartridges etc. We have carried out a Health & Safety risk assessment and are in the process of establishing a written policy document.
b) In your sourcing from suppliers	We are keen to work towards Oxfam's Code of Conduct and will welcome help in doing so.

Contract

At the stage of contract negotiation it is important that environmental considerations are maintained within the finer detail, which can be the first areas to be dispensed at the risk of cost reductions. A contract generally covers two key areas:

(1) The legal terms and conditions of the contract; and
(2) Other conditions specific to the service being provided.

The information provided below is only concerned with the conditions specific to the service being provided. It should also be noted that the basis of the contract should be on a 'partnership approach' in that the service provider can only assist an organisation in reducing sustainability impacts in conjunction with activities carried out by the organisation themselves and any other parties involved. Typical sustainability items in the contract should be constructed from the following three main areas:

(1) Corporate policy – The contract would include the corporate policy and local policy objectives:

As our contractor / service provider you will ...
... Help us to meet our detailed policy objectives (already developed in Step 1: Identification of need).
(Text should be incorporated to explain how the expressed aims of the policy depend on the service forming the subject of the contract.)

(2) Service level agreement – The contract would set out a list of objectives or service level agreements relating to that policy objective:
   As our contractor / service provider you will ...
   ... Detail the specific measures and associated guidance included in the tender requirements / specification, accounting for any modifications made during the negotiation phase. Also detail environmental aspects or work to be carried out.

(3) Monitoring and measurement – The contract would set out how the service provider will be monitored and measured:
   As our contractor / service provider you will be monitored and measured on performance ...
   ... Include a statement that the environmental requirements form part of the basis of performance assessment of the contract (and optionally that unsatisfactory performance would provide equal grounds for termination of the contract).

Reporting

Depending on the environmental service level agreements selected at the specification step and written into the contract, a reporting schedule can be generated. The reporting schedule that is drawn up will be of relevance to both parties in the contract, for use over the contract period.

Management reports will be provided at monthly and quarterly review meetings:

❑ Monthly operational review meetings will be held between the service provider's regional/account manager, or site supervisor and client representative; and
❑ Quarterly strategic review meetings will be held between the contractor's account manager and client and procurement representatives.

Review to include:

❑ Levels of waste disposed, including recycling and recovery rates;
❑ Conformation of compliance with environmental legislation;
❑ Conformance with planned preventative maintenance schedules; and
❑ Management of energy targets.

The results of reporting areas can be communicated via the inclusion of an 'environmental agenda' in regular management meetings between the client and the service provider.

Performance evaluation

The use of performance evaluation will help define and measure progress and achievement against the identified measures. In order to maximise the benefits and ensure cross-referencing, environmental performance should be measured in line with other business performance mechanisms. An example of the balanced scorecard is provided (Figure 2.4) with sustainability inclusions, though your own mechanism should be used where possible.

Take the measures laid out in the service level agreement and for each define:

❑ *Key activities*, i.e. to provide this service the service provider will ...
❑ *Performance level*, i.e. to a performance level which ensures that ...
❑ *Key performance indicators (KPIs)*, i.e. performance is measured by ...

There may be any number of key activities, performance levels or KPIs set for each service being monitored.

In setting performance measures for an individual service, a combination of the different types of measures will be appropriate, to ensure a rounded picture of the service. Many measures, or KPIs, are valuable primarily for on-going monitoring purposes, providing regular information to ensure that a service continues to be delivered to the expected standard. The number should be limited to the three or four that provide greatest impact. However, some measures are set specifically to stimulate improvements in the service. Where this is the case, the required improvements should be expressed as targets. The performance can be attributed by the score and can be evaluated within any existing performance measurement systems.

A score would be attributed to a service provider's performance in order to:

❑ Benchmark performance;
❑ Assess future performance against the benchmark score;
❑ Set a target performance level for the service provider to achieve, i.e. to promote continuous improvement; and
❑ To evaluate any bonus payable.

The measure being assessed must give a binary score, i.e. a 'yes/no' answer. The score attributed may be weighted according to any impacts the particular service may have on the client's business as a whole or on the individual contract, e.g. legal requirements would be given a higher rating. The weightings used for each set of measures should total 100% to give a meaningful result. A negative score may be given for poor performance.

At this stage, performance improvement programmes can be set up to encourage suppliers to improve their performance and support the overall environmental performance of the organisation. The limitations relate to smaller organisations that do not hold the same purchasing power to encourage suppliers. In addition, they require resources in both time and money to be truly effective.

FM ACCOUNT: SERVICE CRITERIA AND PERFORMANCE e.g. Cleaning

Location: _____   Month _____

REF:	SERVICE SUB-ELEMENT	Score[NB]	SOURCE	WEIGHTING	SCORE
001	**Training** Have relevant staff received environmental awareness?	100%		7	7
002	**Waste segregation** Has waste been segregated adequately?	0%		7	0
003	**Products used** Is all COSHH data available?	100%		16	16
004	**Storage** Is all waste stored adequately?	0%		16	0
005	**Incidents** Have any environmental incidents been recorded and investigated?	100%		7	0
006	**Waste documentation** Is all legal waste documentation available?	100%		31	31
007	**Recycling** Have all waste recycling targets been met?	100%		16	16
				Possible	Actual
**NB	TOTAL			100	70

NB: Score 100% if answer=Yes, score 0% if answer=No
**NB: A dark grey score is above 80, light grey score is 70–79

Figure 2.4 Example of a balanced scorecard. Measures quantify a company's performance; targets challenge it to do better.

2.1.6 Existing contractors and suppliers

Aside from the large service provisions such as catering, cleaning or mechanical and electrical, facilities managers will be involved in the tendering of scores of smaller contractor and supplier provisions for the various activities either directly, or through a third-party. The environmental aims and objectives contained within an organisation's policy statement should be communicated and promoted to contractors and suppliers to encourage their support.

With the potential of over 100 contracts in place, there is little value in assessing all of them for environmental impacts and working with them individually. Instead, improved benefits can be gained from identifying those contractors and suppliers that have the greatest impacts on the services provided to an organisation. To assess the level of impact, areas such as business criticality, use or provision of hazardous materials, and annual spend should all be considered. Those identified under any one of these categories should be considered as being a key contractor and supplier, and therefore should be worked with closely to reduce any environmental impacts.

In this way, waste contractors, chiller unit engineers and furniture suppliers may be identified as key suppliers, all of which have a significant impact on the environment. Alternatively, a recruitment agency may also be identified but would have a lower impact, though new temporary employees should be made aware of the practices and processes within the organisation.

It is essential that contractors and suppliers are made aware of, and comply with, relevant requirements and provisions of the environmental processes in place. They should be selected and evaluated on the basis of their environmental performance and competence as well as their technical and commercial proposals.

In order to evaluate all contractors and key suppliers a retrospective review should be performed for existing providers. In addition, new providers which are deemed key should also be made aware of the arrangements at the tender stage. Typical questions to ask during this assessment are:

- ❏ Is a copy of the environmental/sustainability/social/corporate responsibility policy statement (delete as applicable) policy available?
- ❏ Management system implemented, e.g. ISO14001/EMS/SA8000/AA1000?
- ❏ Has a competent person been identified to manage environmental and social issues? If yes, what are the person's qualifications/experience?
- ❏ What general and specific sustainability training is provided?
- ❏ Do you conduct environmental and social risk assessments for your products/activities?
- ❏ Have you had any breaches of legislation or been served with an improvement notice in the past five years?
- ❏ How do you engage with your supply chain on sustainability – provide examples of where this has been effective?
- ❏ What training and education do you provide staff on sustainability? Is this for the supply chain as well?
- ❏ Who will be responsible for sustainability both on-site and off-site?

Suppliers should be able to answer some basic questions:

- ❏ Do you understand and share your ethical and business practices?
- ❏ Have you got the appropriate systems in place to monitor staff knowledge and capability?
- ❏ When did you last update staff on changes to your internal guidance and policies?
- ❏ Are your compensation policies compatible with your ethical practices?

Once a service provider has been awarded a contract, it is important to ensure they are fully aware of the site requirements and individual processes for CR management. This can be provided in a handbook combining CR activities with those of health and safety and/or security issues to form a consolidated document including contact details and what to do in an emergency.

The work to be undertaken, including environmental protection requirements, should be discussed with the contractors, amending job method(s) to be used where necessary to ensure all implications are considered.

The final element is the review and measurement of a service provider's performance over time dependent on the longevity of the contract and its review periods. The company may be subjected to a review at any time, and upon work not being carried out in accordance with an organisation's environmental practices, corrective measures should be the first mechanism set in place and escalated as required.

Any review for environmental activities should be incorporated as part of the standard performance review that takes place, such as service level agreements or business performance reviews (financial, quality, environmental health and safety). In this way, CR practices will gain visibility as an issue important to the organisation, and something that is given a weighting as part of the overall service delivery.

2.2   Environmental legislation

There is an increasing level of environmental legislation that is coming onto the statute books affecting both industrial and office based organisations. Legislation is arriving through a variety of sources including global treaties, country regulations and agreements, which all place common standards on a local, national and international basis. This includes putting the 'producer pays' principle into action for waste products, manufactured items, purchase of property and liability for knowingly causing pollution. This section provides an overview of legislative requirements for specific areas, such as waste management, water management and air and noise control.

Many of the social aspects directly related to facilities management and the workplace are covered by health and safety legislation and hence are not covered in this section. However, wider community based issues, such as nuisance, are included.

Section learning guide

This section provides an overview of the legislative processes and requirements which apply through the various global, regional and local regulatory requirements. It also provides the general position to be applied to minimise risk and manage corporate responsibility impacts:

❏ Overview of legislation affecting the European Union, United Kingdom, United States and Asia Pacific;
❏ Risk management and control affecting air, land and water; and
❏ Role of facilities management in delivering legislative compliance.

Key messages include:

❏ Demonstrable compliance is indicative of a well managed organisation; and
❏ Legislation is increasing, and the requirements to comply are becoming more onerous necessitating greater resources than previously to manage them.

Facilities management has a great responsibility for compliance with many of the legislative requirements noted due to the management and control over issues such as waste disposal, discharge of effluent and monitoring and measurement of plant equipment.

Changes to legislation mean that individuals are liable for environmental damage, similar to that for health and safety incidents. In this way, not only have managing directors been sent to serve a prison sentence, but those lower down the line have been found in breach of environmental legislation. It is the responsibility of everyone within a business to manage environmental legislation and to make relevant people within the organisation aware of unusual occurrences – this includes strange smells from discharges to storm or foul waters, odour from emissions and excessive noise.

In order to manage legal compliance effectively, the applicable legislation must be initially identified and it must be the responsibility of a named individual to identify changes to legislation on an on-going basis.

2.2.1 Overview of environmental legislation in the European Community

Environmental legislation relating to public health, property, town and country planning and criminal law has been in place within Member States prior to the mid-nineteenth century. Since the formation of the European Community (EC), a comprehensive and coherent body of environmental legislation has been developed through a series of directives, which must be implemented in the country and formally recognised by the EC. This has resulted in a considerable amount of the more recent legislation in Member States, being designed specifically to implement EC directives, which may be at odds with older country specific legislation.

Legislation within the EC is moving towards the 'polluter pays' principle, whereby the focus is placed on the producer or manufacturer of products or sub-products to ensure due care is taken with the chemicals' use. It also places responsibility on the manufacturer over the life cycle of that product and its constituent parts. The aim of this principle is to ensure that pollution is prevented at source, therefore minimising its effect during use and disposal.

EC directives give Member States the discretion to determine how the new legislation is to be put into effect: the scope and critical details of the legislation are determined by the directive itself. In some cases, the provisions of a directive may be directly enforceable in the Member States' courts, even where no, or possibly inconsistent, implementing legislation has been enacted in the Member States. This in effect means that:

❑ EC legislation may in some circumstances be directly applicable, whether or not there has been implementing legislation in the Member States; and
❑ If there is a conflict between EC and a Member State's legislation – the EC legislation takes precedence. If it is not clear whether a Member State's statute is in conflict with EC law, a Member State's court has the power to prohibit the government or regulatory authorities from putting the disputed law into effect until the question has been resolved by the European Court of Justice (ECJ).

European community role in legislation

There are four main institutions who manage the identification and implementation of legislation onto the statute books:[11]

The European commission The commission is the EC administrative body that initiates legislation in the form of proposals which are then laid before the council of ministers for discussion and eventual adoption. The council of ministers is the only body which can adopt legislation, although it may, and does in relation to technical and administrative matters, delegate legislative powers to the commission.

The commission executes council decisions and oversees the daily running of EC policies. It also has an express duty to ensure the enforcement of EC law in the Member States, including taking proceedings against them if they fail to implement the provisions of EC law. It is the commission that proposes the areas for new environmental legislation which are set out in its environment action programmes.

The council of ministers The council of ministers is a political body consisting of one minister from each Member State, which meets as and when required. The council is the primary law-making body, and in effect the 'cabinet' of the EC; it reviews, and formally enacts as law, the commission's proposals.

---

[11] NSCA Pollution Handbook 2005, published by National Society for Clean Air and Environmental Protection.

The European parliament The European parliament is mainly a consultative and advisory body with no powers to enact laws although it does have limited supervisory powers over the commission and the budget. It seeks to exercise a degree of democratic control over the running of the EC with the ability to draft legislation and veto measures.

The European court of justice The ECJ has no part to play in the EC legislative process – it is purely the EC judicial body. All ECJ judgements are binding and decisions may, and in some cases do, exert a strong influence on EC policy. The ECJ comprises 13 judges appointed by agreement of the Member States. It is the supreme authority on matters of EC law.

EC legislation

Legislation is initiated through the commission proposing a directive or a regulation, or the council requesting the commission to undertake studies and to submit appropriate legislative proposals. Before the commission issues a proposal for legislation, a consultation and negotiation process takes place formally and informally with interested organisations representing the different sides of industry, consumers, environmentalists and national experts. The commission may issue consultation documents (or 'green papers') setting out its ideas and highlighting problem aspects, perhaps with the draft for a directive.

The commission, after both informal and formal consultation, puts forward a formal proposal for legislation. This will generally be published in the 'C' series of the *Official Journal of the European Union* (OJEU). Once adopted it is contained in the 'L' series of the OJEU.

❏ *Regulations*: Legislative acts that are normally directly applicable in, and binding on, all Member States, without the need for any national implementing legislation. Regulations are applicable from the date when the legislation is laid. Regulations are used when the EC considers that the area to be regulated should offer no discretion to a Member State as to the method of implementation and when complete harmonisation throughout the EC is required. Additionally, the delays that almost always occur if national implementing legislation is needed are avoided. An example is the EC regulation on ozone depleting substances following the Montreal Protocol to ban CFCs and HCFCs.

❏ *Directives*: Require implementation into national law to achieve the aims and results described in the directive. Most EC legislation is issued in the form of directives since they leave the choice of the method of implementation to the member states. Directives commonly require implementation within a stated time, from which the directive has a direct legal effect.

❏ *Decisions*: Legally binding on those to whom they are addressed, whether this is to member states, companies or individuals. They play only a small part in environmental legislation.

2.2.2 Overview of environmental legislation in the United Kingdom

UK legislation is comprised of the recent EC directives and historical legislation. Compliance is achieved primarily through penalties imposed in the criminal courts. Fragmentation within the legislation controlling environmental risks and a low perception of the importance of environmental issues and their impacts to the economic future has resulted in the inadequate infrastructure to manage and prosecute those polluting the environment. Whilst measures are being taken to rectify this, including increased responsibilities given to the Environment Agency (see information box), the level of resources currently available and powers provided to the courts are insufficient deterrents to ensure industry invests in pollution prevention practices.

In the United Kingdom, implementation of directives may be by way of primary legislation (acts of parliament) or secondary legislation by way of statutory instruments or an order under the European Communities Act 1972. Within the UK the use of administrative circulars has been used which is discretionary by the public bodies to whom they are addressed and they cannot usually be challenged effectively in the courts.

However, financial penalties are not far away through fines and prosecutions, and it is advisable for industry to ensure awareness of new and future environmental legislation relevant to it, e.g. draft EC directives. This knowledge should be built into operational requirements to ensure compliance, and if necessary to make legitimate claims for exemptions or amendments.

Environmental liability

A breach of environmental statutory duty can lead to either a criminal or a civil case against the organisation.

Criminal liability

Cases are filed, where a legislative breach can be proved beyond reasonable doubt, based on statutory compliance requirements. In these cases a fine of £20 000 or up to six months imprisonment can be imposed in a magistrate's court. Where a criminal case goes to a crown court, there can be an unlimited fine with up to two years imprisonment. The Environment Agency (or Scottish Environment Protection Agency in Scotland) can only prosecute if they can point directly to a breach of legislation, otherwise it is outside their remit.

Part of criminal liability is to prove the party 'caused or knowingly permitted' pollution:

❑ knowingly means knowledge (mental state) of it;
❑ causing does not have to know; and
❑ it even includes vandalism (put into emergency).

This may occur where there are cases that:

❑ cause statutory nuisance;
❑ deposit unauthorised waste on land;

❏ fail to comply with duty of care, i.e. fail to provide a waste transfer note; or
❏ cause or knowingly permit pollution of controlled waters.

Environment agencies

The Environment Agency for England and Wales and a separate Scottish Environment Protection Agency (SEPA) for Scotland took their powers on 1 April 1996. Part I of the Environment Act 1995 established the agency and transfered the functions, property, power, right and liability of Her Majesty's Inspectorate of Pollution (HMIP), the National Rivers Authority (NRA) and local waste regulation authorities, together with certain functions, property, rights and liabilities of the Secretary of State for the Environment.

The agencies are responsible for protecting and improving the environment in an integrated manner by:

❏ reducing the number of regulators industry will have to deal with and providing a one-stop-shop for environmental regulation;
❏ strengthening integrated approaches to pollution control; and
❏ becoming respected centres of expertise and excellence on environmental monitoring, research and science.

Civil liability

Civil liability can include third-party, nuisance or negligence cases based on proving a balance between two parties and can therefore be more expensive to bring forward than criminal liability proceedings. An organisation could be liable for physical damage, loss of amenity to neighbouring land, disposal of waste onto a third-party's land or causing nuisance from noise, odour or particulates.

2.2.3   Overview of environmental legislation in the United States[12]

In the US, the Environmental Protection Agency's (EPA) primary objective is to protect human health and the environment. To achieve this objective and ensure that decisions are cost-effective and protective, the EPA conducts scientific, economic, and policy analyses. Critically, flexibility is built into regulations from the beginning allowing modification to keep track of changes in science. The process for new legislation involves five main steps:

Step 1: A member of congress proposes a bill. A bill is a document that, if approved, will become law.

---

[12] US Environmental Protection Agency – http://www.epa.gov/epahome/

Step 2: If both houses of congress approve a bill, it goes to the President who has the option to either approve it or veto it. If approved, the new law is called an act, and the text of the act is known as a public statute. Some of the better-known laws relating to the environment are the Clean Air Act, the Clean Water Act, and the Safe Drinking Water Act.

Step 3: Once an act is passed, the House of Representatives standardises the text of the law and publishes it in the United States Code. The US Code is the official record of all federal laws.

Step 4: The law is put into practice through congress authorised government agencies such as the EPA to create the regulations. The EPA researches the need for one, and if required, proposes a regulation for public comment.

Step 5: Following revisions in light of comments, a final rule is published and 'codified' by being published in the Code of Federal Regulations (CFR) – the official record of all regulations created by the federal government.

Key legislation affecting facilities managers in the US includes:

❏ Clean Air Act (CAA);
❏ Solid Waste Disposal Act (SWDA);
❏ Toxic Substances Control Act (TSCA);
❏ Resource Conservation and Recovery Act (RCRA);
❏ Oil Pollution Act (OPA);
❏ Comprehensive Environmental Response, Compensation, and Liability Act Superfund (CERCLA); and
❏ Clean Water Act (CWA).

2.2.4 Overview of environmental legislation in South-East Asia and Australia

Singapore

Singapore has a strong focus on sustainability driven through by the government and implemented through the planning system and the National Environment Agency. Key issues focus on the adjacency of industry to local populations, resource efficiency and best use of land. The government does provide funding, rebates and tax incentives to reduce the impact on the environment, and to start introducing sustainability requirements at the concept stage – before planning. The cost reductions from funding and reduced 'sustainability' costs will more than compensate for any initial start-up costs.

The Environmental Pollution Control Act (EPCA), which came into operation on 1 April 1999, consolidates the previous separate laws on air, water and noise pollution and hazardous substances control. The EPCA therefore provides a comprehensive legislative framework for the control of environmental pollution supporting a range of additional legislation:

❏ Sewerage and Drainage Act (SDA);
❏ Sewerage and Drainage (Trade Effluent) Regulations;
❏ Environmental Pollution Control (Trade Effluent) Regulations;
❏ Environmental Pollution Control (Air Impurities) Regulations 2000;
❏ Environmental Pollution Control (Prohibition on the Use of Open Fires) Order;
❏ European Directive 96/69/EC for passenger cars and light duty vehicles, and 91/542/EEC Stage II for heavy vehicles and European Directive 97/24/EC covering motorcycles/scooters;
❏ Environmental Pollution Control (Hazardous Substances) Regulations;
❏ Environmental Pollution Control (Ozone Depleting Substances) Regulations;
❏ Environmental Pollution Control Act; and
❏ Environmental Public Health (Toxic Industrial Waste) Regulations 1988.

Hong Kong[13]

The Hong Kong Environmental Protection Department (EPD) was created in 1986 as the main government body tackling pollution with the aim of both determining and implementing environmental policy. Responsibilities include:

❏ Planning, developing and managing waste disposal facilities;
❏ Environmental implications of policies and strategic and local plans, and administering the application of the environmental impact assessment process;
❏ Pollution laws and facilitating business to go beyond compliance;
❏ Policy formulation, strategic planning and programme development in the field of waste, water and air management; and
❏ Nature conservation, energy efficiency and community relations aimed at raising environmental awareness.

The main environmental legislation affecting Hong Kong includes:

❏ Air Pollution Control Ordinance (Cap.311);
❏ Noise Control Ordinance (Cap.400);
❏ Waste Disposal Ordinance (Cap.354);
❏ Water Pollution Control Ordinance (Cap.358);
❏ Ozone Layer Protection Ordinance (Cap.403);
❏ Dumping at Sea Ordinance (Cap.466); and
❏ Environmental Impact Assessment Ordinance (Cap.499).

Australia[14]

Environmental legislation in Australia is managed through the Department of the Environment and Heritage (DEH) at government level to focus on matters

---

[13] Hong Kong Environment Protection Department – http://www.epd.gov.hk/epd
[14] Australian Department of Environment and Heritage – http://www.deh.gov.au/

of national environmental significance by:

❑ Advising the Australian government on its policies for protecting the environ-
ment and heritage;
❑ Administering environment and heritage laws;
❑ Managing the Australian government's main environment and heritage pro-
grammes including the $3 billion Natural Heritage Trust;
❑ Implementing an effective response to climate change; and
❑ Representing the Australian government in international environmental agree-
ments related to the environment and Antarctica.

The main legislation affecting FMs in Australia includes:

❑ EPBC Act;
❑ Hazardous Waste Act;
❑ Heritage laws and notices;
❑ Ozone protection and synthetic greenhouse gases legislation and regulations;
❑ Renewable energy (mandatory renewable energy target) legislative framework;
and
❑ Water Efficiency Labelling and Standards Act.

2.2.5   Waste legislation

Waste legislation is underpinned by three principles: precautionary, polluter pays
and proximity. These seek to ensure waste is managed to limit impact on the environment and any reprocessing or disposal takes place close to where the waste is
generated.

Effective waste management is becoming an increasingly important issue for
organisations both from an environmental perspective to reduce waste arisings
and recycling, and from a cost perspective. Part of this is due to the increased
legislation implemented which has increased the cost of removing and disposing
of wastes from site. This rise in cost is likely to continue over the next few years as
legislation drives segregation of waste streams and removal of hazardous materials
to aid recycling and recovery of various materials.

Facilities managers typically manage the various costs and waste streams associated with an organisation from general office waste, confidential waste, fluorescent
tubes or waste oil. The need for greater transparency and segregation of hazardous
and non-hazardous wastes to achieve legislative compliance is likely to increase
costs affecting budgets and the general trend to reduce real estate costs as a whole.

In addition, staff and customer awareness will raise questions on waste management and policy which the FM should be able to answer. These will include areas
such as visibility of waste recycling and recovery, purchase of recycled materials
and knowledge of waste figures. This must be managed alongside the general legislative requirements associated with the storage and removal and final disposal of

the wastes and the provision of suitable waste containers and additional pick-ups from project activities, refurbishments or moves.

The effective management of these various activities by FMs can lead to significant cost savings and service enhancement through proactively working with staff and project teams to identify and provide the capacity to cater for changes in waste volume.

There are three main classifications of waste which affect typical organisations and industry:

❑ Controlled wastes, which include those from commerce and industry. There are separate categories for any mineral, synthetic oil or grease, asbestos or clinical waste;
❑ Hazardous waste, defined under the 14 hazard ratings; and
❑ Inert waste, defined as waste which does not biodegrade or decompose.

Waste is broadly defined as any substance or object which the producer or owner intends to discard. The legal definition is complex, but is worded such that 'wastes intended for recovery and recycling are included within the definition of waste'.

A substance does not cease to be waste as soon as it is transferred for collection, transportation, storage or recovery. Waste is the responsibility of the producer throughout its life cycle regardless of who it is transferred to. The role of FM as either an in-house or contracted service is key to this to ensure compliance is maintained from generation of waste to its final disposal point – the duty of care requires that:

❑ Waste is disposed off in a safe manner;
❑ Only authorised contractors are used to transport the waste;
❑ The disposal site is licensed for the waste (or has an exemption);
❑ Documentation is correctly completed and retained for the statutory period; and
❑ Waste is correctly described and suitably contained.

Duty of care requirements – controlled and inert waste

The legislative requirements require that all reasonable steps must be taken to fulfil the duty of care and complete the relevant paperwork. What is reasonable depends on what you do with the waste. A simple three-step programme should be put in place to ensure compliance:

❑ Step 1 – Make the waste secure;
❑ Step 2 – Give the waste to a registered waste carrier; and
❑ Step 3 – Complete and keep a waste transfer note.

Step 1, make it secure. You must ensure that the escape of waste is under your control by storing the waste safely and securely and preventing it causing pollution or harming anyone. Keep it in a suitable and labelled container protected from vandals, thieves or trespassers or by animals, accident or weather. If you put loose waste in a skip or on a lorry, cover it.

Step 2, if you give waste to someone else, check that they have authority to take it.

Step 3, you must describe the waste in writing. You must fill in and sign a transfer note for it and keep a record of this transfer.

When waste is passed from one person to another, the person receiving it must have an accurate written description of it. The transfer note must be filled in and signed by both parties. Descriptions of the waste can be written on the transfer note. It is not sufficient to write 'general waste' or 'rubbish'. Who provides the transfer note isn't important as long as it contains the right information. Repeated transfers of the same kind of waste between the same parties can be covered by a single agreement. The transfer note to be completed and signed by both persons involved in the transfer must include:

❑ What the waste is and how much there is;
❑ What sort of container it is in;
❑ The time and date the waste was transferred;
❑ Where the transfer took place;
❑ The names and business addresses of both persons involved in the transfer;
❑ Whether the person transferring the waste is the importer or producer of the waste;
❑ Details of which category of authorised person each one is. If the waste is passed to someone for authorised transport purposes, you must say which of those purposes; and
❑ The name and address of any broker involved in the transfer of waste.

Duty of care audit

Legislation governs all aspects of the management and control of waste, particularly in terms of the 'duty of care' placed on the originator to ensure that waste is correctly handled at all stages, right through to its final disposal by a licensed contractor. Organisations must be in a position to demonstrate compliance. Failure to do so will result in fines. Independent and professionally conducted audits of a company's waste management activities will highlight any area where there is a risk of non-compliance and recommend appropriate remedial action.

All waste carriers and end-user sites including landfill sites and recycling centres should be subject to an audit by a competent person. The level of this audit should be based on the risk posed by the type of waste being transferred, either by volume or hazard of waste. This will include the following activities:

❑ Paper trail audit from transfer or consignment notes on site;
❑ Method of material acceptance onto site;
❑ Weighbridge calibration certificates;
❑ Correct licences for company activities;

❑ Up-to-date licences for all sub-contractors and end disposal sites;
❑ Control of discharges to the environment;
❑ Housekeeping;
❑ Training and awareness;
❑ Complaints and breaches of compliance;

Waste electronic and electrical equipment directive

The Waste Electrical and Electronic Equipment (WEEE) directive and its sister document the Restriction of use of certain Hazardous Substances (ROHS) will require companies to review the procurement, management and disposal of electrical equipment. The European Commission estimates an average cost increase of between 1% and 2% for most WEEE products which will affect companies worldwide.

Manufacturers will need to demonstrate their products are ROHS compliant and do not exceed threshold limits for six toxic substances: lead, mercury, hexavalent chromium, polybrominated biphenyls (PBBs), cadmium and polybromided diphenyl ethers (PBDEs). These substances can cause asthma and cancer, as well as damage to the brain, liver, kidneys and the nervous and cardio-vascular systems. In the United States there are already moves to enlarge the list of restricted materials, which can only set the global bar higher for manufacturers wishing to sell in the United States.

The legislation will require producers of WEEE to finance collection arrangements for their products' end-of-life, covered under a series of categories (Table 2.3). Producers also need to include the costs of appropriate treatment and meeting specific targets for recycling and recovery. Producer is defined as any company that manufactures products, re-brands equipment produced by other manufacturers or imports such equipment for sale.

The WEEE directive is based on the concept of 'producer responsibility' – that producers pay for the treatment, recycling and some collection costs of separated WEEE. The directive aims to reduce the overall environmental impacts of waste electrical and electronic equipment, which will help to reduce waste to landfill and improve recyclability. Article 1 of the WEEE directive specifically encourages the re-use of equipment.

The costs of collecting, re-using, recycling and disposing of the equipment and the need to ensure they do not contain banned materials is the responsibility of the producer, and the user to check and verify. As a supplier of components or equipment to any commercial customer, for example an OEM (original equipment manufacturer), compliance to ROHS must be provided to customers, or they risk being removed from supplier lists.

Business-to-business (B2B) transactions must provide for collection, treatment and recycling of old products on the sale of new products. For products placed on the market after 2005, B2B sales must be covered by appropriate contractual arrangements between parties for recovery and recycling. Priority should be given to the repair, upgrade and re-use of whole appliances for original purpose. Collected WEEE must be treated at authorised treatment facilities that meet operational

Table 2.3   Examples of products listed in the directive for each category

Large household appliances	White goods, electric ovens, microwaves, fans, air-conditioning appliances, electric heaters
Small household appliances	Vacuum cleaners, carpet sweepers, sewing machines, irons, electric clocks
IT and telecom equipment	Computers, printers, photocopiers, fax machines, telephones, mobile phones, answering machines
Consumer equipment	Televisions, video recorders, hi-fi, DVDs, electric musical equipment, video recorders
Lighting equipment	Most types of lighting except household light bulbs
Tools	Most DIY and electrical gardening tools
Toys, leisure and sports equipment	Electric trains/car racing sets, video games, coin slot machines, sports equipment with electric components
Medical equipment	Radiotherapy, cardiology, dialysis, ventilators
Monitoring and control instruments	Smoke detectors, thermostats
Automatic dispensers	Vending machines, cash dispensers

requirements. Producers are no longer responsible for all business electrical equipment. They are responsible for accepting WEEE free of charge when supplying new products of equivalent type. This means the cost of any equipment that is disposed off and not replaced, is borne by the business, i.e. the last holder of the WEEE. Business users must separate WEEE from other waste and ensure it is passed for treatment to an appropriate treatment facility.

2.2.6   Water legislation

Water is an essential commodity for life, yet is treated as a free resource used constantly with little thought of reduction or conservation – at least within Europe and North America. Globally, water is a scarce resource with wars waged between nations to recapture control of water supplies, and thousands dying due to dehydration.

Discharges to controlled waters

It is an offence to cause or knowingly permit any poisonous, noxious or polluting matter or any solid waste matter to enter 'controlled waters' – territorial waters, some coastal waters, inland waters and groundwater. The term 'to cause or knowingly permit' has been under debate, with 'causing' generally considered to require a positive act rather than passive inaction, though cause may include the 'siting' of a facility. 'Knowingly permitting' is generally considered to be a failure to prevent pollution and requires knowledge that the polluting activity is taking place.

There is also an exclusion where a discharge may have been the result of an emergency to avoid danger to life and all reasonable steps must have been taken to minimise the effects. An example of this may result from an oil or diesel leak, or from fire water run-off, where prosecutions have previously resulted.

An application for the consent to discharge effluent directly to controlled waters must be made. Where approval is provided, requirements are generally placed to control and minimise the pollution effect from the effluent. Samples are generally required on a regular basis and records must be maintained providing evidence of the testing and calibration taking place and as evidence of compliance with the requirements of the consent.

Discharges to sewers

Discharge of trade effluent directly to the sewer is commonly under the control of sewerage companies. Trade effluent is defined as 'any liquid, either with or without particles of matter in suspension in the liquid, which is wholly or partly produced in the course of any trade or industry carried on at trade premises; and … any such liquid which is so produced in the course of any trade or industry carried on at those premises, but does not include domestic sewage'.

Whilst a sewerage company must accept trade effluent into the public sewer (except where 'red list' substances exist), prior consent must be obtained by the organisation to do so. Once obtained, the sewerage company cannot revoke consent and can only vary its conditions after two years or more from the date of consent or from the date of a previous variation except in certain situations.

Sewerage companies are liable under legislation for the discharge of effluent from their treatment works to controlled waters, and therefore as these requirements change, will alter the consent requirements to organisations. Measurements should be taken at the same specified locations as the local water authority on a similar schedule to provide a check to the invoices received and the quality of the water. Should there be a query, the evidence provided by the check-up samples can be used as proof of compliance with the discharge consent.

Legionella

Legionnaires' disease is a potentially fatal form of pneumonia caused by the *Legionella* bacterium. Breathing in small droplets of water containing the bacterium causes the infection, which therefore cannot be passed from one person to another. The *Legionella* bacterium is widespread in the environment and it may grow in water systems such as cooling towers or hot and cold water systems. The term 'water systems' can include anything from shower/washroom facilities, through to air-conditioning systems and cooling towers. There will be a different level of risk associated with each different water system depending on the type of building and its location.

There is an obligation to consider the risks that *Legionella* bacteria may present to staff or members of the public and to take suitable precautions to prove that:

❏ A competent person has carried out a risk assessment to identify the probability of *Legionella* bacteria existing within your building;
❏ A *Legionella* management programme is available identifying roles, responsibilities and requirements, incorporating regular maintenance and monitoring tasks;
❏ A process or procedure for preventing or minimising the risks identified has been implemented;
❏ Records are kept and reviewed regularly to ensure that they are being properly maintained and completed correctly. Regular reviews take place to assess progress towards fulfilling planned activities and tasks, to identify problem areas, and to set actions for future progress;
❏ The management programme (systems' condition, control procedures and accompanying records/logs) is reviewed by an independent auditor with recommendations incorporated into the following year's plans for action;
❏ Procedures are in place for managing any suspected outbreaks of legionnaires' disease.

2.2.7   Soil and groundwater

Soil and groundwater contamination is becoming an increasingly important area covered by a number of pieces of legislation including the land contamination regulations and the groundwater regulations. Due to the lack of boundaries within the soil and the free movement of groundwater, any contamination that occurs is not only hidden from view, but can traverse a sizeable distance before being contained. The obvious difficulty with this type of pollution incident is to prove knowledge of occurrence and ownership of the pollutant.

Discharges to groundwater

The groundwater regulations affect, particularly for industry, the storage of oil and chemicals from underground tanks, leaking pipes or leaking tanks. The regulations main focus is to:

❏ Prevent the direct or indirect discharge of 'List I' substances to groundwater; and
❏ Control pollution of groundwater resulting from the direct or indirect discharge of 'List II' substances.

Oil pollution

The majority of oil pollution incidents occur because of a lack of basic precautions, such as placing bunds around oil tanks and fitting alarms to warn of overfilling. These simple measures are minimal in cost compared to the pollution clean-up costs and penalties.

The aim of any oil regulations is to significantly reduce the number of oil related incidents through establishing design standards for all above-ground oil stores and secondary containment, such as a bund or drip tray, to be in place to prevent oil pollution. The regulations apply to all external oil containers with a capacity of greater than 200 litres, including fixed tanks, intermediate bulk containers (IBCs) or drums. The capacity of a tank is the maximum volume that the container can hold, rather than the amount of oil usually held within the tank. Therefore, if a 300 litres capacity tank is never more than half full, it will still be captured by the requirements of the regulations.

All oil containers are required to have secondary containments, such as a bund or drip tray, with sufficient capacity to retain 110% of the container's contents. If there is more than one tank within the bund, then the bund must hold 110% of the largest tank or 25% of the aggregate, whichever is the greater. Other provisions include the requirement to ensure that the bund and container are impervious to oil and water and to ensure that any penetration of the bund, e.g. by the fill point, is effectively sealed. The provisions for fixed tanks include the positioning of pipes, leak detection and overfill prevention devices and pumps. Valves must be situated within the secondary containment and the fabric of the container should be impervious to oil and water. Finally lids must be kept on all drums stored outside.

Underground tanks

Underground storage tanks also pose a similar risk, although the length of time to identify a pollution incident will be longer. The same requirements on pollution prevention apply, banning the discharge of petrol, diesel and other toxic chemicals into soil and groundwater. In order to safeguard against leakages, similar provisions should be made to underground tanks as for overground tanks – alarms for overfill or leaks, double skinned tanks, controls over leakages when filling. In addition, regular checks on joints and integrity of the tank are recommended to ensure ongoing compliance with the regulations.

Contaminated land

The legislation surrounding contaminated land is an attempt to regulate historical contamination and associated potential liabilities. The requirements for contaminated land operate under the rule of 'caveat emptor', or buyer beware, is still applicable, with the due diligence process required to identify any potential contamination either directly on site, or in the local area.

The definition of contaminated has been defined as 'any land which appears to be in such a condition, by reason of substances in, on or under the land, that'

(a) Significant harm is being caused or there is a significant possibility of such harm being caused; or
(b) Pollution of controlled waters is being, or is likely to be, caused.

The legislation looks to assess sites; including identification of pollutant linkages and presence of sources, pathways and targets to identify the relationships between the substances and its environmental receptors. Owners and landlords need to be aware of potential liabilities of not only direct contamination to the site but also indirect contamination, e.g. creeping pollution, from adjacent land.

The potential liabilities to those managing buildings are:

❏ *Owner* – Liable for past poor practices with regard to site operations of others such as by the previous owner or tenant who cannot now be traced. Also liable for contamination which has occurred since ownership.
❏ *Tenant* – For historical pollution, if the original polluter cannot be found, the liability will rest with the landlord or owner of the building. However, tenants should ensure that their activities do not pose a risk to the environment to avoid the threat of future liability.

Liability directive

Businesses primarily affected by the new European directive are those involved in traditionally-polluting activities, such as plants releasing heavy metals into water or into the air, installations producing dangerous chemicals, landfill sites and incineration plants. Certain industries including oil and nuclear will fall outside of the directive's scope and will continue to be covered by their own liability regimes. However, liability may fall on the directors and officers of construction companies and other site operators. Moreover, in regard to contaminated land and statutory nuisance, the owner or occupier may be held liable for environmental harm.

A controversial aspect of the proposal, at least as far as industry is concerned, is the wide definition of 'environmental damage'. Not only does it cover land and water pollution but also damage to the biodiversity of any protected species or habitat. It creates a financial liability in regard to any measure required for the clean-up and restoration of contaminated land that poses a potential threat to human health. For owners and occupiers of land that was once contaminated, it would be advisable to check the current health of your property to be assured it continues to be free of contaminants.

2.2.8 Air and noise legislation

There are a number of controls on air pollution, with the IPPC (Integrated Pollution Prevention and Control) directive being the main piece for industrial premises. It was developed to provide an integrated approach to the management and regulation of certain industrial processes through the use of best available techniques to reduce the environmental damage to land, water or air. Other industrial processes predominantly come under legislative requirements for clean air and statutory nuisance for air quality as a whole. The presence of ozone-depleting substances is the other key environmental hazard which has come under strict legislative pressures through direct legislation from the EC.

Aside from the health and safety aspects of noise at work, the legislative requirements of noise relate predominantly to the nuisance providing a defence for local residents and communities against noisy industrial or office based neighbours.

Office locations are not exempt from noise issues, where a loose bearing on an extractor fan can cause a nuisance to local residents at night time if not fixed.

Flue gases

Boilers should be maintained to their maximum efficiency through a planned preventative maintenance programme and the use of an appropriate computerised system, where appropriate, to enable the boilers to burn at their most effective rate. This provides not only energy savings but minimises the exhaust of combustion gases. A contingency plan should also be developed to provide a back-up, should the system fail.

Volatile organic compound emissions

Emissions of volatile organic compounds (VOCs) are generally very low due to the small quantities used. This specifically concerns the use of solvents in paints, thinners and cleaning chemicals. Alternatives to solvent based products are fully available and should be used as a substitute. This will also mean these chemicals do not necessarily have to be stored in flammable proof cupboards dependent on their hazardous nature.

Project air pollution

It is important that any dust or particulate matter that may be generated as part of a project to refurbish or construct a building is carefully controlled and managed. Any release into the atmosphere will impact upon local residents and the community, which can be acted on. During the design phases of the project, if the generation of dust is identified as a major impact, the local community should be included in the communication loop to gain their buy-in at this stage of the project and make them aware of the potential for releases to occur.

Noise pollution

The greatest potential for noise pollution is from the re-construction of many parts to the building fabric across the site. Each project, similar to the requirement for air pollution, should be risk assessed for various aspects, including noise and recommendations made to minimise potential pollution. This includes specifying working hours, use of equipment or soundproofing.

Ozone depleting substances control and management

Bans exist on the supply and use of chloroflurocarbons (CFCs), halons and other ozone depleting substances (ODSs) from 1 January 2001.

❏ 'Supply' is defined as change of ownership, even without payment. It is referred to as 'placing on the market' in the regulation;

❑ 'Use' is defined as use in the production, maintenance or servicing of equipment. Running an existing CFC appliance, without maintenance, would *not* qualify as use.

The use of HCFCs (hydrochlorofluorocarbons) has been prohibited for the manufacture of new equipment in all refrigeration and air-conditioning applications since 1 January 2001. There will be a ban on the use of virgin HCFCs from 1 January 2010 and a ban on the use of all HCFCs, including recycled materials, from 1 January 2015.

All ODSs used in refrigeration and air-conditioning equipment must be recovered during servicing and maintenance of equipment, or prior to dismantling or disposal. Recovered CFCs must now be destroyed by an approved process. Recovered HCFCs can either be destroyed or can be re-used until 2015.

All practical precautionary measures must be taken in order to prevent and minimise leakage. An important new rule is that fixed equipment containing ODSs, which has a refrigerating fluid charge greater than 3 kg, must be checked annually for leakage.

ODSs Affected by the Legislation

Refrigerants	Solvents
CFCs: *11*, *12*, 13, 113, 114, 500, *502*, 503	CFC: 113
HCFCs: *22*, 123, 124	1,1,1 trichloroethane
HCFC blends: various including R401a,	HCFCs: 141b
R402 a, R403a, R406a, R408a, R411b	CBM (bromochloromethane)
Common trade names: Arcton, Forane,	Common trade names: Arklone,
Freon, Isceon, Solkane, Suva	Freon, Flugene, Forane, Kaltron,
	Gensolv, Genklene
Foam blowing agents	Fire-fighting fluids
HCFCs: 22, 141b, 142b	Halons: 1211, 1301

Note: some of the trade names quoted above are used only for ODS substances whilst others are used for both ODS and non-ODS substances.

Italic numbers refer to compounds most commonly used.

2.2.9 Energy legislation

Changes in energy legislation are providing a key driver for facilities managers to incorporate energy efficiency objectives into their project plans. The importance of this cannot be understated because as electricity and gas prices have continued to rise there has been limited legislation within this area.

EU energy performance of buildings directive

The EU energy performance of buildings directive (EPBD) will have far-reaching implications for the owners, operators and developers of all buildings across Europe (both domestic and non-domestic) and will play a vital role in delivering the Kyoto Agreement's targets and energy-efficiency objectives. The EPBD provides a major opportunity to achieve the step-change in buildings-related energy efficiency.

Key provisions of the directive are:

❏ Minimum requirements for the energy performance of all new buildings;
❏ Minimum requirements for the energy performance of large existing buildings subject to major renovation;
❏ Energy certification of all buildings (with frequently visited buildings providing public services being required to prominently display the energy certificate); and
❏ Regular mandatory inspection of boilers and air-conditioning systems in buildings.

The Dutch government has commented that the 'high administrative costs' of implementing the directive precluded its implementation 'in the short term' with an estimated cost of Euro 80 million per year.

A key issue is the EPBD requirement that whenever a building with a total useful floor area of over 1000 m² undergoes major renovation, its energy performance is upgraded to incorporate all cost-effective energy-efficiency measures. The directive also requires that all new buildings should meet minimum energy performance requirements. For those with a useful floor area over 1000 m² governments must ensure that, before construction starts, formal consideration is given to alternative systems for heating.

The EPBD requires that whenever a building is constructed, sold or rented out, a certificate (no older than 10 years) detailing its energy performance must be made available by the owner to the prospective buyer or tenant. In order to facilitate comparisons between buildings, the energy performance certificate must include reference values such as current legal standards and benchmarks. It also must include recommendations for the cost-effective investments which can be undertaken in the building, and which will improve its energy performance. All buildings, either occupied by a public authority, or regularly visited by a large number of people, must display in a prominent place clearly visible to the public its current energy certificate.

The introduction of building energy certification is likely to have a profound effect on the commercial property sector. No organisation which has any concern about brand equity, or its corporate social responsibility standing is going to want to occupy a poorly rated building – particularly if environmental reporting or pension disclosure requirements result in naming and shaming.

The EPBD provides member states with two options for reducing the energy consumption of boilers. The first option is to lay down the necessary measures to establish a regular inspection of boilers fired by non-renewable liquid or solid fuel of an effective rated output over 20 kW. The second option is for governments to ensure that there is adequate provision of advice to users on the replacement of the boilers, other modifications to the heating system and on alternative solutions, which may include assessment of the efficiency and appropriate size of the boiler.

In order to reduce energy consumption of air-conditioning systems, governments must establish regular inspections of all air-conditioning plant with an effective rated output of more than 12 kW. Such an inspection must include an assessment of the efficiency and sizing of the air conditioning plant, compared to the cooling requirements of the building. Appropriate advice must be provided to users on possible improvements or replacements, and on alternative solutions.

EU member states must ensure that certification of buildings, the drafting of the accompanying recommendations and the inspection of boilers and air-conditioning systems are carried out in an independent manner. This must be by qualified and/or accredited experts. These can operate as sole traders or be employed by public or private bodies. If a member state can demonstrate that there are an insufficient number of qualified or accredited experts anywhere within the European Union to implement fully the provisions associated with building certification (or plant inspection), they may delay introduction for up to three years. If they wish to cause this delay, governments must justify this to the commission together with a schedule, detailing precisely when they do plan to fully implement the directive.

2.3 Business case and corporate challenges

The role of facilities and property management and its relationship with an organisation's strategy has historically been negative where the role of property has not been considered in the decision-making process. The challenge for those involved in the property and FM industry is how to raise the profile of the industry to the strategic level. The growth of CR, coupled with stakeholder requirements and raising legislative fiscal requirements has placed it firmly at the highest level.

Section learning guide

This section describes from a corporate perspective the drivers for CR and provides the basis to enable the facilities manager to develop an effective business case to deliver corporate responsibility aligned with and supportive of business needs:

❑ Drivers for business and organisations from a global and local perspective;
❑ The main benefits and drawbacks to the implementation at a business level;
❑ How to develop an effective business case and management programme; and
❑ Implementing the business case and delivering improvement.

Key messages include:

❑ Non-tangible benefits, including staff retention, employee satisfaction and stakeholder relations, often are greater than direct cost returns; and
❑ Support from the top is critical for success.

2.3.1 Drivers for global corporate responsibility

There has been intense public reaction against companies that are seen as over-powerful, those that produce over priced goods at the expense of deprived countries or that ruin the environment. Shell's pollution of land and deprivation of the local communities in Nigeria, Nike's sweatshops in the Far East and Nestlé's production and distribution of milk powder to newborn babies are well known. Less known are the dramatic effects these scandals had on share price and the 'saleability' of the brand name.

It is clear that many multi-nationals already understand the business risks of being perceived as environmentally unfriendly, or as a 'bad' citizen, and the substantial benefits of being a good one. And there is ample evidence to show it is equally important for smaller organisations to ensure they too demonstrate responsible behaviour, even at a very local level.

Organisations see benefits in using CR as a vehicle to provide governance, which can help them to respond to the following fundamental changes:

❏ Whilst globalisation has created new opportunities, it has also increased the complexity and fragmentation of business activities abroad. This has resulted in greater risks and management requirements on a global scale, particularly in developing countries.
❏ Brand image and reputation are key facets to an organisation's role in a competitive business environment. Consumers and stakeholders are increasingly requesting information about the conditions and sustainability impact in which products and services are generated. Where organisations are found lacking, consumers will move away to socially and environmentally responsible firms.
❏ A greater role by the investment banks has increased the disclosure of information beyond traditional financial reporting. Organisations providing more robust data are rewarded with greater investment.
❏ As knowledge and innovation become increasingly important for competitiveness, enterprises have a higher interest in retaining highly skilled and competent personnel.

The challenges to further take-up and adoption of CR practices are:[15]

❏ Knowledge and an understanding about the relationship between CR and business performance to develop a business case;
❏ Consensus between the various parties involved on an adequate concept taking account of the global dimension of CR, in particular the diversity in domestic policy frameworks in the world;
❏ Teaching and training about the role of CR, especially in commercial and management schools;
❏ Awareness and resources among SMEs;

---

[15] Green Paper Promoting a European framework for Corporate Social Responsibility – http://ec.europa.eu/employment_social/soc-dial/csr/greenpaper.htm

❏ Transparency, which stems from the lack of generally accepted instruments to design, manage and communicate CR policies;

❏ Consumers' and investors' recognition and endorsement of CR behaviours; and

❏ Coherence in public policies.

The adoption of CR into the policies and practices of an organisation is based on the business case which underpins this, and ultimately the risk to the business for not following this course. This is based on the risk to the image and stakeholder pressure, particularly consumers. Ultimately, to take CR forward there must be the buy-in from industry and organisations to provide sufficient knowledge and resources. However, the true aim is to match the benefit to an organisation by raising the level of social development and providing greater value in society through local authorities and other NGOs. This can and should only be performed in partnership with the government strategy to ensure the longevity of such activities and the support from key members within the community.

Socially responsible investment (SRI)

With the aim of pushing the use of SRI issues, recent legislation has required pension funds to disclose whether and how they take account of social, environmental and ethical factors in their investment decisions. This has had two major effects, firstly as pension funds start asking for more information from organisations they invest in, the advantage in publicising information on social and environmental performance is recognised. Secondly, those industries and individuals involved in developing the rating have created a series of factors and requirements to quantify the performance by organisations, providing league tables of compliance.

2.3.2   Drivers for corporate responsibility at a local level

Corporate responsibility is presenting new challenges to all areas of business and development. Within the built environment, sustainability is affecting not only how buildings are designed and constructed, but also how they are operated, occupied and managed on a day-to-day basis.

There are a range of drivers affecting business and stakeholders including the following:

❏ International institutions and treaties;

❏ Legislation and regulation;

❏ Markets;

❏ Business standards;

❏ Risk management;

❏ Campaigning organisations (NGOs, charities, pressure groups, business groups);

❏ General public opinion.

Economic instruments are being used across a range of markets, taking a variety of forms – taxes, trading schemes, subsidies, regulation, voluntary agreements and information schemes – and often these instruments work alongside each other.

2.3.3   Key benefits and drawbacks

The last few decades have been marked by numerous changes in the regulatory framework relating to protecting the environment and promoting a shift towards sustainable development. Sustainability represents a management framework that drives us to seek continuous improvements, in a way that integrates economic, environmental and social objectives into both our daily business decisions and future planning activities. It also represents an approach for unlocking opportunities for improving sectoral competitiveness and enhancing reputation. Such an approach can bring clear benefits to business by helping to:[16]

Reduce operating costs

❑ Using passive, climate-sensitive systems correctly can cut heating and cooling energy consumption by 60% and lighting energy requirements by 50% and yield a good rate of return based on the initial investment;
❑ Water-efficient appliances and management of them can reduce water consumption by up to 30% or more;
❑ Cost savings can be achieved by minimising waste across the supply chain and across all stages of development, leading to higher material efficiency and the greater use of recycled materials;
❑ Site-clearing costs can be lowered, by minimising site disruption and movement of earth and installation of artificial systems;
❑ Earnings are possible from sales of re-usable items removed during building moves or demolition;
❑ Energy efficiency and optimal energy-use strategies during the occupancy of a building can result in energy bills being reduced and a reduction in greenhouse gas emissions, leading to reduced liabilities for owners and occupiers and meeting future compliance requirements; and
❑ Using local suppliers and maximising fleet efficiency can lower transportation costs.

Reduce capital costs

❑ Rehabilitating an existing building and re-using existing built assets can lower infrastructure and materials costs;
❑ Highly serviced buildings cost more to build and use more energy and materials;
❑ Energy-efficient buildings can reduce their equipment needs, with significant cost savings;
❑ Improved efficiency in the use of energy and natural resources has been shown to save 10% of waste at no cost as well as reducing the impact on the environment; and

---

[16] Approach for unlocking opportunities, improving sectoral competitiveness and enhancing reputation – http://www.bre.co.uk and http://www.envirowise.gov.uk

❏ Sustainability investment decisions made early on in a project are likely to result in less of an increase in capital costs than those made at a late stage and may result in significant life cycle savings.

Reduce liability and avoid risk

❏ Liabilities are resulting in larger compensation awards so providing increased incentives to minimise these at the outset can save major costs;
❏ In an increasingly litigious society, concerns over liabilities are increasing and can provide an incentive to move away from traditional solutions;
❏ Insurers are becoming increasingly concerned about the effects of climate change such as storm damage and flooding, and this is being reflected in high premiums and more stringent conditions attached to payouts; and
❏ Fines for water, land, air and noise pollution incidents can be avoided through more effective environmental management systems.

Enhance reputation and attract investors

❏ Sustainable buildings can create an enhanced corporate image, market reputation and shareholder value;
❏ Investing in sustainable buildings can serve to avoid reputation damage due to adverse publicity;
❏ As the perceived risks of stock ownership are reduced, shareholder value increases; and
❏ Investment appeal can be influenced by attractiveness of a building project or company. Improving ratings with investors and insurers can lead to a lower cost of capital.

Contribute to and maintain competitive advantage and differentiation

❏ In a highly competitive sector with low margins, the concept of 'sustainable building management' is going to become an increasingly important differentiator;
❏ Gaining a reputation as a leader in sustainability can differentiate a company and deliver a competitive advantage over other companies not yet addressing these issues; and
❏ Offering a sustainable building can help to attract and retain good clients/tenants.

Provide future proof and manage change

❏ Building investors and occupiers are increasingly concerned about future proofing their business and investments against future changes in legislation;
❏ There is a major risk in not identifying and keeping up-to-date with future legislation and regulation of relevance to business;

❏ With the average life of a procurement programme lasting several years the current rapid rate of change in legislative requirements makes this a significant issue; and
❏ By designing structures and systems that are flexible and adaptable to other uses, costs of reorganisation and renovation are reduced.

Manage the supply chain

❏ It is important that each stage of the construction process (from inception through planning, design, build, operation, occupation and demolition) is considered as part of the complete cycle and within an overall framework;
❏ Partnering, whether formalised or simply through improved relationships, can improve this integrated view. Sustainability can only be integrated in a planned and well managed way, through prioritising different aspects of sustainability according to the needs and resources of specific projects;
❏ Stronger long-term relationships and ways of working through greater use of partnering arrangements can produce cost savings of up to 50% and time savings of up to 80%;
❏ Building partnerships with suppliers increases trust, with regular contracts resulting in competitive pricing; and
❏ Sustainable organisations are likely to be more attractive to those clients with corporate responsibility policies, or that offer preferential bid status to companies with good sustainability management policies and practices.

Favour productivity and workplace welfare

❏ Operational resource use costs account for approximately five times construction costs over the typical 60 year life of a building. Typical staff costs account for approximately 200 times the construction costs over the same period;
❏ The financial benefit of a 5% increase in productivity is roughly equivalent to the full property costs of a typical organisation. Studies have shown that improved indoor environments with maximum daylighting, natural ventilation, and indoor air quality, can increase employee productivity by up to 16%; and
❏ Investment in sustainable buildings can improve an organisation's ability to attract and retain the best staff. Employees in buildings with healthy interiors have less absenteeism and longer retention. Staff contentment is vital in achieving productivity, image and bottom-line financial benefits. Ensuring healthy indoor air can reduce insurance and operating costs and reduce liability risks.

Generate opportunities for new business and emerging markets

❏ Sustainable building expertise and technologies are central elements in the emerging environmental industry sector and are potential export products. Recycling creates jobs. Diverting these materials to local processors instead of local landfills creates new economic opportunities.

Deliver community benefits

❏ By demonstrating a thorough understanding and responsiveness to customer and local community concerns a company can help to maintain good 'community relations';
❏ Sustainable buildings can help to support and protect the local economy through stimulating demand for local building materials, jobs and industries. They can also reduce the demand on – and the need for – public infrastructure, such as power plants, water supply, landfills, sewer systems and waste-water treatment plants. More efficient building occupation means less money leaving the locality; and
❏ Sustainable buildings have a positive effect on the surrounding community by preserving wildlife habitats and green spaces and avoiding noise and pollution. Sustainable buildings typically emit 40% less CO_2 and help to mitigate climate change.

Despite the many positive trends and drivers, there are also some practical factors working against sustainable development reporting. In particular:

Cost versus benefits

A company may incur significant costs, both in terms of human and financial resources, for producing a sustainable development report. The cost will be considerably less if the data is already being collated for specific business purposes.

Systems versus reports

A long lead time is necessary to create data gathering systems for new parameters. This explains why companies tend to report on issues that they are familiar with, as reporting on new topics requires the development of dedicated systems to collect the necessary information. This can prove both very expensive and time-consuming.

Transparency versus legal implications

Some companies are wary about the uses to which their sustainable development information will be put. How do they deal with environmental and social information that leaves room for interpretation and potentially misinterpretation? Others are reluctant to publicise their adherence to codes of conduct as they fear the future legal implications and lawsuits this may entail.

2.3.4   Building blocks for making a business case

Investing resources in enhancing reputation can increase and sustain shareholder value. Done well, sustainable development reporting can demonstrate to stakeholders that the company is honestly striving to meet stakeholder performance values and expectations across financial, environmental and social dimensions.

The market

Sustainable development is best achieved through markets that encourage innovation and efficiency. Markets are human constructs and we need to continue to improve them to best serve the needs of society if we want to maintain global open markets.

The right frame

If basic framework conditions push us all in the wrong direction, then that is the way society will go. Business needs governments to set appropriate framework conditions which can support its efforts to move toward sustainable development.

Eco-efficiency

A main business contribution to sustainable development is eco-efficiency. By adopting eco-efficiency measures, a company will improve both its environmental performance and its financial results.

Corporate responsibility

Corporate responsibility is an evolving concept that is always being redefined to serve different needs and times. This leads to a constant debate about the respective roles of government and business in providing social, educational and health services. Also, how far along the supply chain does a company's responsibility extend?

Learning to change

Corporate concern for sustainability requires change throughout the corporation. A sustainable business excels on the traditional financial return but it also embraces environmental performance, and community and stakeholder issues.

From dialogue to partnerships

We need to go beyond stakeholder dialogues toward partnerships that combine skills and provide access to constituencies that one partner may not have. They also improve the credibility of the conclusions and actions.

Informing and providing consumer choice

If business believes in a free market where people have choices, companies must accept responsibility for informing consumers about the social and environmental effects of those choices.

Innovation

To become more sustainable, companies must innovate, that is, continuously modify or invent new products, services, and manufacturing processes that are more eco-efficient than their predecessors. However, unless companies engage stake-holders in their innovation processes, they will not succeed in gaining social or marketplace acceptance.

Reflecting the worth of earth

We do not protect what we do not value. Proper valuation will help us use the markets to maintain the diversity of species, habitats, and ecosystems, conserve natural resources, preserve the integrity of natural cycles, and prevent the build-up of toxic substances in the environment.

2.3.5   Developing a business case

The main drivers for developing a business case for adopting the processes and practices of CR should be based on four key mechanisms (Figure 2.5):

(1) Company structure and model and whether the existing structure would best facilitate the development of these processes. This will also include a review of the resource requirements to achieve adoption (Figure 2.6);
(2) Drivers – the main drivers relate to the management of business risk including client base, customers, internal processes, brand and values and stakeholders in general. In addition the benefits of greater PR can often provide greater visibility and provide more effective sales;

Group CR vision and principles

- Identify key stakeholders and their needs
- Identify and express company vision and principles
- Undertake a CR review to define baselines and understand the gaps

Consistent strategy implementation across business units

- Develop a strategy for vision implementation
- Establish CR goals and resources
- Agree KPIs to meet management and wider stakeholder requirements
- Implement consistently across business units against plan

Tailored KPIs, targets and effective information flows

- Establish information collection systems
- Establish framework for effective knowledge sharing (internal and external benchmarking)
- Internal assurance of data quality and consistency
- Preparation of CR reports

Identification and exploitation of value creating opportunities

- Communicate assured company CR performance to stakeholders
- Demonstrate links between CR management and business performance
- Identify areas for innovation and improvement

Figure 2.5   Developing a business strategy.

Sustainable development

Corporate strategy
-Vision
-Values
-Strategic planning

Current and future activities

Figure 2.6   Corporate strategy.

(3) Business strategy is particularly important in the medium term to identity and focus upon the critical corporate responsibility issues affecting them. It is important to identify and quantify the indirect benefits which may arise, and the longer term benefits where an organisation has its greatest influence; and

(4) Impacts to be identified which will be based on the risks the process is trying to manage. Ultimately, the greatest use of any new system is to change the internal culture, as opposed to implementing only the cost-effective and visible measures.

Although most businesses support the assumption of a positive impact of CR on competitiveness, particularly in the long term, they are however not able to quantify this effect.

Why does your company need a CR policy?

If you have a CR policy then your organisation is likely to be viewed more favourably by the market place and this policy can be highlighted in your marketing and public relations activity. This will help to attract and retain high quality staff and also portray a positive image to the local community and shareholders. Demand from the financial community for information on the environmental performance and liabilities of organisations are becoming more commonplace. Increasingly, this information is having an impact on share prices and the way these organisations are viewed by customers, shareholders, staff, investors and regulatory bodies.

Investments are being made in those organisations that are seen to manage in a sustainable and ethical manner. The trend is to exclude those companies who do not have an effective environmental and sustainable management policy. This could affect the £3.5 billion of funds currently invested in the FTSE 100 companies.

In July 2001, a new FTSE index was unveiled – FTSE4Good – similar to the Dow Jones Sustainability Index (DJSI). This index only includes companies that have been recognised as:

❏ Being socially responsible, based on environmental sustainability;
❏ Operating a positive stakeholder relations policy; and
❏ Supporting human rights.

For the first time investors will be able to identify those who fall short of these exacting sustainable management requirements and make their investment decisions accordingly.

Evidence has shown that firms who manage their environmental risks also manage their business effectively. Thus, expenditure for both management systems and pollution prevention can be justified on a purely financial basis – they can lower a firm's cost of capital by reducing risk. With increases in environmental taxation, coupled with the rising costs of environmental liabilities, the business case for sustainable management will continue to grow.

2.3.6   Implementing CR into FM

CR will impact upon a great number of an organisation's business activities including FM, and therefore it is important that clarification is gained at the outset of the main goals and vision for the resulting process. As a new entity, to fully understand the requirements can take a significant level of resources and time. This can be exacerbated where an organisation operates not only in different countries, but where cultures and social roles are markedly different. Whilst the diversity helps to create innovation, it can also pose a great challenge in providing any kind of uniformity across the global portfolio through a visible structure.

Ultimately, the need to adopt these practices is best incorporated through service provision such as cleaning or catering roles, and the activities performed by contractors and suppliers. However, the elements of CR are far less understood than environmental management and therefore need greater definition at the local level to ensure the due governance and robust nature expected.

Developing the vision

The first element in providing CR is to develop a clear concept and understanding of what it means to the organisation, and the values which it defines. The role of CR, similar to ISO14001, goes over and beyond the requirements of legal compliance, involving areas such as equal opportunities, health and safety, social responsibility, local community and environmental management.

This will involve a major culture change at the senior management level to:

(1) Recognise that its activities have a wider impact on the society in which it operates;
(2) Take into account the economic, social, environmental and human rights impact of its activities across the world; and
(3) Seek to achieve benefits by working in partnership with other groups and organisations.

There is today a growing perception among FMs and their wider counterparts that sustainable business success and shareholder value cannot be achieved solely through maximising short-term profits, but instead through market-oriented yet responsible behaviour. In this context, an increasing number of organisations with the support of FM have embraced a culture of CR. Despite the wide spectrum of approaches, there is a consensus on its main features:

(1) CR is behaviour by businesses over and above legal requirements, voluntarily adopted because businesses deem it to be in their long-term interest;
(2) CR is intrinsically linked to the concept of sustainable development: businesses need to integrate the economic, social and environmental impact in their operations; and
(3) CR is not an optional 'add-on' to business core activities – but about the way in which businesses are managed and the culture in how they operate.

For the building management chain, this approach means working together to promote a more sustainable environment (with regard to the environmental, social and economic aspects of work being carried out). The service provider can only assist the client to reduce impacts in conjunction with activities carried out by clients themselves and any other parties involved. Therefore, it should be noted that the basis of these documents is to promote a 'partnership approach'.

The approach in this section deals with the environmental and social dimensions – there will also be themes that will be relevant to health and safety.

Partnership approach

It appears that companies are increasingly more willing to accept that engagement (where companies encourage investors and other parties involved to improve their social, ethical and environmental performance) is not a vehicle to unsurp management's roles and responsibilities: rather they are seeing it as a beneficial process that can highlight potential issues before they become a serious problem.

Source: Co-operative insurance / Forum for the Future

Good management of sustainability issues offers many strategic benefits to companies. A company's property forms a significant part of a company's environmental and social impacts. The benefits of sustainable building management are set out in many documents. A particularly relevant document:[17] 'Reputation, Risk and Reward' (Sustainable Construction Task Force) sets out the benefits to companies in the property sector. These include:

❏ Avoiding reputation damage and risk due to adverse publicity over performance on environmental, social and business probity issues;
❏ Differentiating a company and delivering a competitive advantage over other companies not yet addressing these issues;
❏ Improving ratings with some investors leading to a lower cost of capital and insurance;

---

[17] *Reputation, Risk and Reward*, A report by the UK Sustainable Construction Task Force – http://projects.bre.co.uk/rrr/index.html

❑ Assisting in meeting compliance requirements and reducing the risks of not identifying and keeping up-to-date with future legislation and regulation of relevance to business;

❑ Improving relationships with suppliers, contractors, sub-contractors and associated professions; and

❑ Better ratings in indices such as DJSI or FTSE4Good leading to a preferential status on bids.

Cost / benefit of sustainability

There are a number of operational benefits, including cost savings from greater material efficiency, energy efficiency and landfill disposal; avoiding prosecution for water, land, air and noise pollution incidents; demonstrating best practice. The FM company has the ability to assess the cost/benefit of each measure and implement all those that meet the criteria defined by the client. The cost/benefit analysis can be on any basis the client decides but would ideally combine a financial assessment (payback or rate of return) and qualitative sustainability elements.

Transparency is a key element of the CR debate as it helps businesses to improve their practices and behaviour; transparency also enables businesses and third parties to measure their results. CR benchmarks against which the social and environmental performance of businesses can be measured and compared are useful to provide transparency and facilitate an effective and credible benchmarking. The interest in benchmarks has resulted in an increase of guidelines, principles and codes during the last decade, though they do not provide the need for effective transparency about business, social and environmental performance. CR benchmarks should build upon core values.

Development and implementation of a policy statement

The reasoning behind the development of a policy statement is to provide transparency and visibility to the entire process, ensure the robustness through communication to all parties and provide a mechanism to co-ordinate activities. The policy is also the top level document which will be seen by the customer, stakeholder and NGO and be brought into question for the various comments contained within.

Development of a policy statement or code or practice describes how an organisation relates to key risks affecting it, such as labour issues, human rights and the environment. This is particularly important where organisations are transboundary and have varying environmental and social requirements. Also included should be a statement that all applicable legislation will be met as a minimum, with an aim to implement the standards across all locations. The biggest challenge related to codes is to ensure that they are effectively implemented, monitored and verified.

Policy should include appropriate mechanisms for evaluation and verification of their implementation, as well as a system of compliance and involve the social partners and other relevant stakeholder which are affected by them, including those in developing countries, in their elaboration, implementation as well as monitoring. In many ways this is similar to those statements currently provided through the existing environmental policies. In keeping with other policy statements, the development of this document must be supported and backed with a management system which enables confirmation of the various elements and visibility of progress and actions taken.

Management standards

Once a policy statement has been developed, the various contents of it must be implemented into day-to-day practices. This will involve integrating an environmentally and socially sound culture within the management and financial structure of an organisation. Since there is no clear mechanism to achieve this, it is recommended that the policy be implemented in one of three ways:

(1) Incorporating into the various aspects and impacts into the ISO14001 structure to develop the organisations objectives and targets (cost parameters, risk to business, liability, impact on local community/bio-diversity, etc). Whilst not a specific requirement the risk identification and management cycle can be followed. The process for implementing a management system is detailed in Section 2.4.

   This process can be implemented with minimal cost and interference to business practices and is likely to be well received and backed by staff. However, it can take much longer to develop and implement than going through the ecological footprint route.

(2) The eco-management and audit scheme (EMAS) allows organisations to evaluate, manage and improve their environmental and economic performance based on the ISO14001 requirements. The EMAS system requires the inclusion and review of stakeholders and community activities where there is an influence by the organisation.

   This will include active employee involvement as a driving force for EMAS alongside the contribution to the social management of organisations. The development of a report specific to the stakeholder will also focus an organisation on the activities it must progress to maintain community acceptance.

(3) Ecological footprinting is a method of measuring the impact that an organisation has upon the environment, community and through the supply chain (Section 1.2). It calculates the influences and impacts an organisation has based on the day-to-day practices performed, including use of utilities and resources, procurement of goods and services and waste generated and disposed. The effects on the local communities are also measured from both a positive and negative perspective.

   These results can then be benchmarked to other similar organisations and targets set for improvement. To implement this system a large amount of baseline data is required to ensure that the footprint is accurate and the targets set are appropriate for the organisation.

The first element of implementing a management system is the identification of the legislation requirements and the risks attached to the FM business activities

Table 2.4   Risks identified for FM activities

Positive	Negative
Increased staff loyalty	Increased staff costs
Improved customer satisfaction and development of partnership role	Poor staff perception and development of 'contractor' role
Reduced staff (organisation and contractor) turnover	High turnover of staff leading to increased costs and poor continuity, stifling creativity
Procurement of sustainably sourced goods, e.g. food, furniture	Procurement of hardwoods originating from non-regenerating rainforest
Provision of disabled access to meet or exceed legal requirements	Exclusion of personnel due to physical attributes

performed. This will include not only the health and safety aspects and environmental areas covered already, but ethical and training management practices likely to be implemented by an organisation through investors in people (IiP). Effectively, this review should highlight all the risks inherent within the FM role including impacts by other departments (as customers) and stakeholders such as the local community from which many cleaning, catering or security staff are employed.

The types of risks typical for FM related activities are linked to employee or contractor welfare, use of materials and chemicals (Table 2.4). In a similar way to managing the significant risks identified as part of the environmental or health and safety systems, CR risks should be controlled where possible through documentation, objectives and targets or continued monitoring and measurement.

Following the identification of the risks, one of the control mechanisms is to ensure adequate documentation is provided to cover all aspects of the business risks, how they are to be managed and where necessary further operational documentation to closely control practices. Again, much of this can be factored into existing health, safety or environmental documentation. Where possible, baseline figures or benchmarks should be identified to provide some kind of relationship between recorded and expected or best practice figures. These should be developed from the major risks identified and the commitment documented within the policy statement.

Suitable training and awareness should be provided to all staff and other stakeholders to gain their buy-in. This will include communication made to shareholders, customers, contractors and suppliers and the local community.

Measurement, reporting and assurance

Through the development and assessment of the various risks, there will be a need to provide regular monitoring and measurement to deliver the verification and visibility of the policy statement. To provide the greatest transparency a report should be written detailing the vision and the steps taken to achieve this vision. There has been a rapid increase in the public reporting of CR indicators. Initially many of these reports appeared as sections within annual reports. However, the setting up of a number of awards has raised the perception and knowledge of reporting and placed the development of a CR report as industry practice (Section 2.5). Due to the unstructured nature of CR, any report should provide for flexibility specific to the requirements of FM activities.

If required, external certification can be provided as a level of assurance through a number of recognised bodies. Based on the ISO14001 or EMAS management system implementation, certification can be achieved through these traditional routes, though this will not provide evidence of the socially responsible element implemented. An alternative is to obtain alternative certification and qualification from a number of organisations that prescribe standards for business practices such as Good Corporation, SA 8000 or ACCA.

2.4   Corporate responsibility management systems

Over the past five years organisations are increasingly requesting assurances from their contractors and suppliers for environmental practices and governance in business activities. This has led to the rise in developing and implementing environmental management systems, of which FM has a major, and in many cases, a leading role. Coupled with this has been the increased recognition of the social aspects surrounding the local community, employees and contract staff.

Section learning guide

This section reviews the growth in implementing environmental and social management systems, the business benefits and reasoning behind this, as well as providing an overview of setting a corporate responsibility management system (CRMS) in place:

❏ Introduction to management systems, how they operate and how to get the best from them;
❏ Definitions and distinctions between environmental, social and corporate responsibility management systems;
❏ Advantages and disadvantages of management systems; and
❏ Implementation of a social and environmental management system.

Key messages include:

❏ The best management system is incorporated into the business culture and structure to become part of day-to-day life;
❏ Simple is best – communication and training over documentation; and
❏ Resources during the initial review and risk assessments to fully understand the issues and mechanisms to manage them are critical to the success of any system.

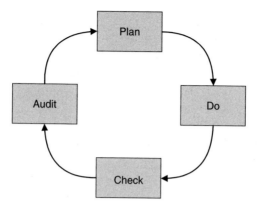

Figure 2.7   Deming model for management systems.

2.4.1   Introduction to management systems

The concept of systems to provide structured continuous improvement came from work with Japanese companies in the 1950s performed by W. Edwards Deming (Figure 2.7). The structure was about learning from your successes and failures in a systematic manner, understanding the reasoning and building these lessons back into the process. Deming devised the Plan-Do-Check-Audit structure which has been the mainstay for the modern-day management system.[18]

In the 1970s and 1980s many European and American organisations took this model and implemented variations into their strategic and operational functions. Total quality management (TQM) was based on the systematic process to provide continuous feedback and re-engineering of services based on knowledge learnt throughout the life cycle of the product or service.

Management systems have long been used to identify, plan, transfer and monitor knowledge throughout an organisation or within a role. Systems have been developed, initially focused on quality, to ensure effective and controlled management of this knowledge to provide business benefits and improved results. Many of these steps, such as setting goals, targets, evaluation and review, are basic elements of any good management system. Many companies have utilised this structure to develop quality and health and safety systems based on these ideals.

Environmental management pre-1990 was largely focused towards ensuring legal compliance and minimising the potential for major incidents to occur. The process was predominantly based on the reduction of environmental litigation which was perceived as a competitive disadvantage. From the early 1990s the social consciousness of organisations started to be questioned by shareholders driven by a group of non-governmental organisations (NGOs) playing on the role of the organisation within the community in which they operate – whether in the immediate locality, in-country or globally.

---

[18] Deming Cycle – http://www.balancedscorecard.org/bkgd/pdca.html

Over the past ten years there has been a major shift change in the perception organisations have on the environment and the society they operate within, which has been discussed in Section 1.2. This has led increasingly to the development of management systems incorporating environmental and social criteria based on the Deming model.

2.4.2   Good practice management systems

Personal experience has seen a number of management systems which do not deliver environmental benefits resulting in wasted resources, inefficient systems and in the worst case prosecutable offences not picked up through the management system.

An MEPI (measuring environmental performance of industry) project surveyed 270 firms and 430 sites across Europe and found that there was no positive relationship between certified EMS and improved performance.[19] It also found that fossil-fuel-based energy producers showed a negative correlation – their performance deteriorated. Furthermore, the Policy Studies Institute has found that enforcement action is just as likely at sites with EMS as without and that waste sites with EMS perform worse than non-EMS waste sites.[20] Some sites just want the badge on the wall, so they can stay in business and in the supply chain. Effectively they are just buying the certificate – it's tokenism. Clearly, there is a need to differentiate between the businesses that take their responsibilities seriously and those that don't.

It is also suggested that management systems may lead to complacency by some companies that may believe monitoring data is enough, and don't act upon it. The key thing is whether you hardwire the management system into the company's cost accounting system. Critically, if there is an obvious and established link between cost savings and environmental impact then it is far more likely that management systems will deliver environmental improvements. Companies that recognise the fundamental link between CR and good business management have good and effective management systems.

So, although there may be no direct link between management systems and CR improvements, it is the corporate culture that determines whether or not companies take their responsibilities seriously. Management systems can help them make meaningful strategic decisions, but in itself they will not guarantee better performance.

Regulatory bodies are increasingly using the development of an EMS and certification to help an organisation understand its impact on the environment, stimulate sustainable practices and ensure they comply with legislation. The justification is based upon the premise that a management system will provide visible evidence to the regulator of compliance and therefore make it easier to not only audit but also provide confidence in legal compliance.

---

[19] MEPI (Measuring Environmental Performance of Industry) project – http://www.sussex.ac.uk/Units/spru/mepi/about/index.php

[20] Policy Studies Institute – http://www.psi.org.uk

The development of corporate responsibility management systems (CRMS) has come about from the growth of separate social and environmental management systems – largely because there is no standard sustainability version available to achieve certification.

To fully contribute to improved performance, a good management system should:

❑ Be implemented at a strategic level and integrated into corporate plans, and policies. Top-level commitment is required so that senior management understand their role in ensuring success;
❑ Identify the organisation's impacts on the environment and society and set clear objectives and targets to improve their management of these aspects as well as the organisation's overall environmental performance;
❑ Be designed to deliver and manage compliance with laws and regulations on an ongoing basis, and will quickly instigate corrective and preventative action in cases of legal non-compliance;
❑ Deliver good resource management and financial benefits; and
❑ Incorporate assured performance metrics that demonstrate the improvements and value delivered by the management system.

2.4.3   What is an environmental management system?

There are four main types of environmental management system (EMS) that an organisation can implement. The major differences between the types of EMS are relatively simple, though there are scores of minor variants, particularly for the latter two systems that can both achieve certification. However, for all four systems the end aim is continual improvement and the prevention of pollution:

❑ An internal 'home-grown' EMS without certification;
❑ Step-by-step systems to develop an EMS, e.g. BS 8555;
❑ Certification to BS EN ISO 14001: 2004; and
❑ Certification to eco-management audit scheme (EMAS).

Home-grown EMS

Home-grown solutions are commonly cheaper to develop and implement than certificated systems, due mainly to the implementation of relevant parts of the system and external certification fees not being required. Whilst these systems enable an organisation to focus on facets of the business which will benefit the most, this comes as a consequence of not providing a rounded approach and therefore potentially leaves certain areas devoid of the same focus. It is recommended that any home-grown system is based on the same model as ISO 14001 to ensure the foundations of a solid risk assessment and legal compliance check is made.

How this is achieved provides the greatest benefit of not following a certified system directly. Two of the main approaches of developing a bespoke EMS are through specific environmental data monitoring, such as energy, waste, water and

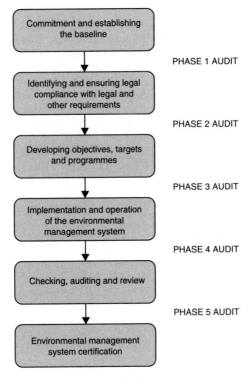

Figure 2.8 Step-by-step EMS development.

transport information or through a scoring roadmap. Both of these options are particularly applicable for multi-site operations.

Step-by-step EMS

A growing range of environmental management system certification is being developed focused on the small and medium size enterprise (SME). Typically these organisations do not have the capital or resources available to invest in the time and money required to implement a typical EMS. This new breed of system enables development to take place in bite-sized chunks slowly building up to a full EMS with the potential of certification. Typical examples which exist are BS 8555,[21] which outlines an implementation process that can be undertaken in up to six separate phases and allows for phased acknowledgement of progress towards full EMS implementation. This has particular reference for small businesses, although it is applicable to any organisation, regardless of the nature of the business activity undertaken, location or level of maturity.

The flowchart (Figure 2.8) describes the typical step-by-step process for the implementation of an EMS with a phased audit programme at the end of each step.

---

[21] BS8555: Guide to the Phased Implementation of an Environmental Management System from the British Standards Institution (BSI).

In this way, an organisation may take a gap between the closure of one audit and the commencement of the next activity. BS 8555 will help all businesses to improve their environmental performance and will demonstrate to interested parties that progress is being made towards the target level of environmental management. Increasingly environmental criteria are being set in contract tenders and even existing clients may start to ask suppliers for proof of their environmental credentials. Soon it may become less a case of 'can small businesses afford to put in place an environmental management system' and more a case of 'can they afford not to?'

Certification to ISO 14001

The most widely recognised and internationally respected EMS is the ISO 14001 standard which provides a common approach regardless of country, activity and size. It specifies a framework of control for an environmental management system against which an organisation can be certified by an external body to a 'standard' that looks for continual improvement through the identification and control of environmental impacts in line with the Deming model. The standard itself is a short and succinct document written in a legalistic manner containing approximately 88 different elements which must be covered to the satisfaction of the 'auditor' to gain the certificate.

ISO 14001 forms part of the ISO 14000 series – a number of documents focused upon providing a standardised methodology to delivering environmental practices within business. The 14000 series covers a range of documents to support organisations to identify, manage, audit and reduce their environmental risks. The box provides a list of the current documents which are constantly being reviewed.

ISO 14000 series documentation

- ❏ ISO 14004 provides guidance on the development and implementation of environmental management systems;
- ❏ ISO 14010 provides general principles of environmental auditing (now superseded by ISO 19011);
- ❏ ISO 14011 provides specific guidance on auditing an environmental management system (now superseded by ISO 19011);
- ❏ ISO 14012 provides guidance on qualification criteria for environmental auditors and lead auditors (now superseded by ISO 19011);
- ❏ ISO 14013/5 provides audit program review and assessment material;
- ❏ ISO 14020; 14021; and 14024 provides information on environmental labelling and declaration issues;
- ❏ ISO 14030+ provides guidance on performance targets and monitoring within an environmental management system;
- ❏ ISO 14040; 14041; 14042; and 14043 covers life cycle issues;
- ❏ ISO 14050 provides environmental terms and definitions;
- ❏ ISO 14060.

The ISO 14001 Standard is focused upon internal environmental issues, which do not require the view of local community activities or global actions to be taken into account for most cases. For many, whilst the rise of ISO 14001 has highlighted that good environmental practice can be a benefit to organisations, it has missed the opportunity of providing real benefit on a global scale to re-address the procurement of goods and services. That is not to say that these issues cannot be included within a certificated system, but their inclusion within an organisation's aspects and impacts is not a requirement of the standard. To provide real change it is recommended incorporating these procurement issues within the system, particularly where products are bought from outside the European Union or United States, such as furniture. The system is most common in the United Kingdom and Japan, though is gaining popularity across the globe.

External certification means that organisations can demonstrate to shareholders, regulators and the public that their system has been audited, in the same way as are their financial accounts, by those with appropriate professional skills, and knowledge. The information provided by a certified system will be seen as being more credible. Other benefits of external certification include:

❏ Confidence that the system meets recognised requirements and standards;
❏ Enhanced value and assurance to customers in the supply chain;
❏ Independent review of the way the organisation is committed to its activities and their associated impacts on the environment;
❏ Closer involvement of employees and other stakeholders; and
❏ Protection of reputational value.

Certification to EMAS

The revised EMAS which was re-launched in early 2001 now uses ISO 14001 as the initial step which makes it far more accessible and meaningful to the range of organisations with an ISO 14001 certification. EMAS allows a targeted approach to report development with tailored information for stakeholder groups, e.g. regulators, financial institutions, suppliers, customers and neighbours. It is recommended that dialogue takes place prior to producing these reports which can also be produced in electronic format, e.g. CD-ROM or web page, or paper format.

EMAS has not caught on within the United Kingdom, with relatively few private organisations seeking certification to this level. It has been highlighted as a means to implement Agenda 21 for local authorities and has a wide take-up. Within Germany and Japan the number of business choosing EMAS over other schemes is very high, commonly outstripping those who gain certification to ISO 14001. A major reason is the need for reporting, which has far greater visibility and is expected within the German and Japanese marketplace.

2.4.4 What is a social management system?

Unlike the EMS standard ISO 14001, there is no internationally recognised standard for the development and implementation of a certified social management

Figure 2.9 AA 1000 social management system framework (adapted from AccountAbility).

system. The two most widely recognised are AccountAbility's AA 1000 and Social Accountability (SA) 8000, both of which follow the same intent as the Deming cycle.

SA 8000 is applicable to all organisations regardless of size, country and sector and therefore is relatively generic and wide ranging to take into account the scope of business it must cover. It covers areas affecting employees and related labour laws, supplier and contractor working practices, working hours and remuneration, and health and safety practices.[22]

The AA 1000 framework is designed to improve accountability and performance by learning through stakeholder engagement (Figure 2.9). It was developed to address the need for organisations to integrate their stakeholder engagement processes into daily activities and has been widely used by a variety of groups across the globe. The framework helps users to establish a systematic stakeholder engagement process that generates the indicators, targets, and reporting systems needed to ensure its effectiveness in overall organisational performance. The principle underpinning AA 1000 is inclusivity. The building blocks of the process framework are planning, accounting and auditing and reporting. It does not prescribe what should be reported on but rather the 'how'. In this way it is designed to complement the *GRI Reporting Guidelines*.[23,24]

The AA 1000 series is based on three propositions:

(1) Stakeholder engagement is key;

[22] Social Accountability International – http://www.sa-intl.org/

[23] AccountAbility – http://www.accountability.org.uk/aa1000/

[24] Global Reporting Initiative – http://www.globalreporting.org/

(2) Accountability is about the ability for an organisation to act on this engagement;
(3) Organisations must learn and innovate effectively on the basis of stakeholder engagement.

Over 100 organisations have implemented the AA 1000 standard partially or wholly including a number of major organisations covering a wide range of sectors – Westpac Banking, Vodafone, Unilever, Sydney Water, SABMiller, Novo Nordisk, Nike, Coca-Cola and BHP Billiton. In addition a number of government practices have implemented the standard, predominantly from Australia, Italy, Canada and the United Kingdom. It is particularly noticeable that a large proportion of organisations are from Australia where the promotion of greater transparency, involvement and inclusivity has been heavily promoted by industry, the community and government as a whole.

2.4.5 Developing a sustainability management system

As yet, there are no international standards for a sustainability management system (SMS) which have been recognised and accepted within the marketplace. The lack of a standard has led to a range of alternative systems being put in place which are partially contradictory in their approach and delivery. It is anticipated that an internally recognised sustainability management system will be developed by the ISO group shortly. Until the development of a standardised methodology, there are three main ways of implementing a system:

❑ An internal 'home-grown' SMS without certification;
❑ Joint certifications to an environmental and social management system, e.g. ISO 14001 and AA 1000;
❑ Implementation to a sustainability management system, e.g. SIGMA.

Home-grown SMS

As with the development of an environmental management system, as described above, an internally developed sustainability system will provide much the same benefits as well as the same difficulties. The ability to focus upon key areas will enable the system to become more flexible in meeting the internal, customer and stakeholder requests. The system can be increased in terms of its scope covering social and environmental impacts as it matures and grows.

Joint certification system

To meet with increasing requirements to demonstrate environmental and social practices, the current option is to gain certification to joint management systems – integrating together the environmental and social aspects together alongside the additional corporate requirements into a single integrated management system. The development of the integrated system will follow the same process as any other management system. The difficulty will lie in identifying an appropriate external certification body that is competent to audit the system.

The reason for this difficulty lies in the approach of the external certification process. Each certification must meet a set of minimum criteria set from both the environmental and social systems, which an auditor can assess against. However the ethos behind the certified systems is different and may lead to conflicts between how the auditor views the system and how the organisation wishes to implement it.

Sustainability management system

The SIGMA Project – *Sustainability – Integrated Guidelines for Management*[25] – was launched in 1999 supported by the UK government to provide a systematic framework to enable organisations to be more sustainable. More recently the British Standards Institute has released BS 8900: *Guidance for Managing Sustainable Development*, which provides a business perspective to implement CR.[26]

2.4.6   Advantages and disadvantages

Some of the real benefits, particularly external recognition of your efforts via certification, can only be gained though a structured approach to environmental and social management. In addition a more formal approach will increase commitment across the company, facilitate on-going improvements and identify further cost savings on an on-going basis. As certification to a management system is voluntary, companies find it harder to justify spending time and money on implementation, rather than simply meeting legislative requirements.

The key advantages are:

❑ Improving cost control from tighter management control over environmental and social impacts including raw materials, waste, energy and community issues;
❑ Staying within the law, and identifying new and impending legislation to maximise benefits;
❑ ISO 14001 like ISO 9001 is becoming a 'market requirement' in many industries;
❑ 'Being seen to be corporately responsible';
❑ Reduced insurance premiums;
❑ Can be integrated with existing management systems such as ISO 9001 or health and safety systems with which it has common elements;
❑ Avoiding the risk of costly prosecution by keeping up-to-date with the environmental legislation governing the business's activities;
❑ Achieving significant cost savings through improved efficiencies in areas such as water, energy and raw materials;
❑ Improving environmental performance and reducing the risk of pollution incidents and associated liability costs;

---

[25] Sustainability – Integrated Guidelines for Management Project – http://www.projectsigma.com
[26] British Standards Institution Sustainability Standard – http://www.bsi-global.com/environmental/sustainability/bs8900

❏ Offering access to new business opportunities where an EMS is a requirement;
❏ Enhancing credibility with customers, stakeholders and the general public;
❏ Providing evidence of sound management; and
❏ Boosting staff morale and improving their awareness of their environmental responsibilities.

The disadvantages are:

❏ The time and money that will have to be spent on the development and maintenance of the system, in the short-term;
❏ Difficulties in integrating with current management systems;
❏ Diversion from dealing with other key issues;
❏ If competitors do not implement ISO 14001 you may be less competitive in the short-term; and
❏ Reducing the flexibility of management.

What you need to know

❏ Not having a management system may damage your business. Many major organisations are already dictating that suppliers must secure the certification. It's worth talking it through with your clients and customers;
❏ It helps cut risks and costs. Implementing it should help you identify the risks you currently – and perhaps unknowingly – face. It could also save you costs as you refine your operational processes;
❏ You can do it bit by bit. You don't have to put your whole operation through the hoops; the standard can be implemented site by site – or even unit by unit;
❏ It offers short-cuts in tenders. If you secure certification you will be able to get automatic short-cuts when it comes to certain tender processes;
❏ Pressure on polluters is mounting. With increasing fines from enforcement agencies, the standard acts as a safeguard, cutting the risk of penalties;
❏ Call in the experts for outside help. It's a complex process which requires particular expertise. Calling in consultants may be a good way of hiring skills and bring fresh eyes into your operation. Plus it will allow your own team to carry on with its own work; and
❏ You need dedication. Implementing a management system will take time and cost money. With staff training a vital element, all staff involved from top to bottom must be 100% behind it.

Small businesses are responsible for an estimated 60% of the commercial waste in England and Wales and for as much as 80% of pollution incidents. Limited time and resources mean that many smaller businesses ignore environmental issues to concentrate on core business, but in doing so they could be risking costly prosecution and missing out on commercial benefits that have a real impact on the bottom line. A survey by the UK Environment Agency of more than 8000 small UK businesses last year – SME-environment 2003 – found that only 18% could name any environmental legislation and just 23% had implemented any practical measures aimed at reducing their environmental impact. Only 4% of the businesses we surveyed had an environmental management system in place, reflecting the European Commission's recent report of pitifully low interest in EMSs among small businesses. More encouragingly, 41% of businesses we asked said they wanted more help and advice on green issues.[27]

To support the business a free website has been developed to provide clear guidance on the environmental legislation affecting a range of business sectors, as well as good practice advice on issues such as waste management and water and energy efficiencies. However, the website is only part of the bigger picture. To change environmental performance in the long-term it is crucial that environmental issues become central to the overall management processes of a business, rather than merely an afterthought.

2.4.7   Costs and timescales for implementation

Companies are finding the major environmental management system development cost is employee time. ISO requires that all employees are informed about the company's environmental policy, and specialised knowledge among those whose job may have a significant environmental impact. Training programmes can be resource intensive in time lost from production plus costs of instruction, and it takes time to document procedures for certain critical operations. Companies with high employee turnover or multiple sites may find knowledge transfer intranet/extranet technologies and self-directed learning tools effective to develop and maintain employee capabilities and minimise some of these costs.

The key to resourcing a system is to link together with the existing system to reduce the workload and additional activities for staff. Training and education can be provided as a top-up to the overall system, rather than for the individual requirements of the new management system. The aim is to provide a system which is not seen as onerous or a bolt-on – if this is how it is perceived, there will be little benefit.

Capital costs of environmental management system development are relatively small in comparison, assuming companies already have appropriate control equipment and monitoring instrumentation in place to be in compliance with federal, state, and local requirements.

The costs and benefits of environmental management systems can be difficult to measure. How do you measure the value of a preventive system? The costs incurred in the course of complying with regulations such as monitoring and permit requirements are potentially hidden among other items, such as overhead accounts. Intangibles such as enhanced consumer response indeed have value, but that value may depend on how your company is currently positioned on these issues and its goals.

---

[27] Environment Agency Report – http://www.netregs.gov.uk/netregs/1169119/

The development of a management system is a lengthy process that can take an operation employing 2000 or so people up to two years to complete. The common timescales are between 9 and 24 months depending on the resources (man-hours and capital) and commitment available at the beginning. This will also be based on the level of management system being implemented with an in-house system requiring less time than a certifiable system.

Environmental and social management systems are predominantly based on a two-tier external audited programme, with the stage one audit reviewing compliance of the documented system against the requirements of the standard. It is recommended this takes place once the documentation has been written and partially implemented with feedback available from staff to gauge the level of implementation required. This will enable modifications to be made relatively easily to the management system without having implemented a significant level of time and resources.

In addition, the external certification body will nominate the same lead auditor to operate both the first and second audits and therefore there is a level of continuity and assurance over any changes which do need to be made. However, if you feel the lead auditor has not met expectations from failure to understand the business or being excessively pedantic on issues, you have the opportunity and ability to change who this individual is. This change has happened many times and does not impact or affect the certification process other than needing to bring a new lead auditor up to speed with the system being implemented.

The second stage of implementation must take place following a management review and sufficient time for some issues to have been resolved – commonly three months. An effective management review will take into account the past twelve months of operation, results from audits performed, training needs and the overall performance. Together, gaps will be identified and measures to close out the issues identified and planned. Closure of the planned targets against timescales will provide confirmation to the certification body that sufficient resource and commitment is available to maintain the system.

The costs taken to implement an environmental/social management system should be seen in relationship to the efficiencies, cost and resource savings and competitive advantage that can be achieved. It is recommended the following resources be provided for a typical organisation (Table 2.5).

As with many implementation projects, the greater the level of work performed at the outset, the more robust the system will be and therefore greater benefits derived. This also relates to the level of commitment and resources required to maintain the system.

2.4.8 Developing and implementing a corporate responsibility management system (CRMS)

The role of facilities management in an environmental/social management system is critical due to the nature of the operations and activities performed. FM

Table 2.5   Typical resource requirements for the development of a management system

Resource	Full-time equivalent throughout project (%)	Resource	Cost
Environment manager	30	External audit	Approx. £550/day
Environmental assistant	40	Training	£5000
Steering group members	5	Magazines	£700
Working committee	20	Awareness	£1000

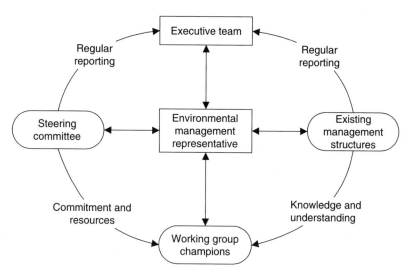

Figure 2.10   Environmental/social management structure.

traditionally operates the 'dirty' activities including waste management, utility management, plant maintenance, refurbishments, cleaning, catering and security.

Within these are the major visible environmental and social impacts and traditionally the main areas for objectives and targets such as reductions in energy consumption and waste disposal. The role of FM should encompass the management of these effects on the environment and society and the achievement of targets. FM should play a key part in a steering group and working committee in the implementation of an environmental/social management system (Figure 2.10).

The development of an environmental/social management system is a time-consuming activity that is best performed utilising a knowledgeable resource as a focal point, known as the sustainability management representative (SMR). This role will co-ordinate activities, maintain timescales and the project plan and ensure pinchpoints or deficiencies are highlighted prior to issues occurring. The SMR will chair and be supported by a steering committee comprised of senior management members from the main functional business units. The role of this committee will be to ensure the resources, commitment and co-ordination with business activities is provided to aid the implementation and change of culture.

The main work will be performed by the larger working group which will include champions from across all business units including areas of finance, human resources and sales and marketing since each of these have environmental impacts and can provide joined-up thinking for the various processes.

All of this activity must be provided in conjunction with existing management structures and business processes to ensure additional work is not needlessly generated, but also so that all the business risks affecting each impact is managed.

It is important to incorporate the system into pre-existing structures and method-ologies to ensure that there is a culture fit between how an organisation performs and the new corporate responsibility management system (CRMS). Without this fit, the CRMS can be seen as being an additional piece of work and a bolt-on to the day-to-day practices. Whilst this will not necessarily mean that the CRMS will not be implemented, it will certainly take additional time, effort and resources to achieve the same goals throughout the implementation process and for on-going management.

What are the steps in a corporate responsibility management system (CRMS)?

An implemented CRMS, either as a home-grown system or by gaining certification, is a fairly flexible standard which not only makes an organisation look good to the outside world, but one which brings a host of business benefits. The process to developing and implementing a system has been described in the 17 steps outlined below.

The roles provided by facilities management are detailed below within each of the various sections, based on a supporting role being provided in the imple-mentation of the CRMS. Where FM plays a greater role, further activities may be performed.

(1) Identifying those parts of your operations that have significant impact on the environment and society. This will include ensuring a robust significance criteria is developed to assess each of the environmental aspects identified based on the activities, services and products of the organisation (more detail on this process is provided later in this section). This process is the cornerstone to the development of an SMS and therefore no shortcuts should be made in this activity. Typical environ-mental and social impacts will include utility management, waste management, chemical control, control of suppliers and contractors, project management and churn activity, planned preventative maintenance and inspections, and statutory requirements.

Identifying environmental and social risks

As mentioned previously, the cornerstone to an effective environmental and social management system is the identification of environmental and social risks (aspects and impacts or cause and affect) and its assessment to identify significance. A FM who has the greatest impact on the environment should take where possible a proactive approach to effectively identify and manage these risks to minimise their impacts.

The risks associated with the activities, products and services of an organisation should be considered from the following areas involving the inputs and outputs from the processes on site. This should include those which could have an impact, positive or negative, on the environment and those which could affect the local community.

❑ Sustainability policy;
❑ Regulatory, legislative and other requirements;
❑ Exemptions to legislatory requirements;
❑ Manager and employee knowledge;
❑ Existing environmental and social practices and procedures; and
❑ Emergency situations.

This should include normal considerations, start-up and shut-down and impacts associated with potential emergency situations. Those impacts, which are not directly controlled or influenced by an organisation, should be eliminated from the list. The aim of the impacts should focus on where an organisation has the ability to influence and change practices. Where this is not achievable there is little point in including the area as an impact.

The process used to characterise and evaluate environmental risks should be carefully considered to ensure the findings provide a basis for prioritising aspects according to the severity of their potential impact upon the environment. In order to achieve this it is beneficial to use a quantitative assessment process as opposed to a subjective yes-no scoring process based on legal requirement or policy requirement.

The quantitative assessment is based on the likelihood of the environmental risk occurring multiplied by the severity of the risk to the environment, which is based on areas such as legislation, volume, storage controls, business risk and cost to rectify (Figure 2.11). Both are scored on a one to five scale (one as low and five as high) giving a total score of 25. Any risks which are a legal compliance requirement should be deemed significant automatically.

A decision should be made on the significance score for which above this score the risk is significant and therefore must be controlled and below is not significant. The table in Figure 2.11 provides an indicative risk assessment where a score of ten was deemed significant, with risks scoring up to 16 a medium risk (light shading) and above 16 a high risk (dark shading).

For each of the registered significant aspects, control measures should be identified, documented and implemented in order to prevent or reduce the impact of that risk upon the environment. A single control procedure may relate to more than one aspect and similarly a single aspect may be covered by more than one control procedure.

(2) Identifying what relevant laws and regulations affect your organisation, what effect they have and the criteria required to meet and maintain compliance.

OCCURRENCE ↓					
5	5	10	15	20	25
4	4	8	12	16	20
3	3	6	9	12	15
2	2	4	6	8	10
1	1	2	3	4	5
SEVERITY →	1	2	3	4	5

Figure 2.11 Quantitative risk assessment methodology.

This will also include the obligations that an organisation subscribes to through memberships and alliances. An overview to environmental legislation has been provided in Section 2.2 with more specific details to implement good practice provided in Chapter 4.

(3) Developing a corporate responsibility policy 'statement', based on the significant impacts and legal compliance requirements, that should incorporate the objectives to prevent pollution and meet legal compliance. The policy document should be signed by the site director or managing director of the business and made freely available. If necessary, an environmental policy can be designed specific to a business activity such as FM or real estate processes. In this case the environmental risks and legal compliance requirements specific to FM or real estate would need to be identified and assessed, with a policy statement signed by the head of the area.

Developing a corporate responsibility policy statement

By making [a] Corporate Commitment, you are making a public demonstration of your organisation's values, vision and sense of responsibility and showing your stakeholders how serious you are about improving your efficiency and investing in the future.

www.MACC2.org.uk

The policy statement should be a short and succinct document from which the rest of the system will support and deliver. By its nature, the statement will provide the main objectives and briefly how you intend to deliver these. The following section describes the three main elements of a policy statement.

Recognition of impacts We recognise that everything that we do has both direct and indirect adverse impacts on the environment and society.

Overall policy aims

❑ Reduce our direct adverse impacts wherever we have managerial control;
❑ Influence others to reduce our indirect impacts;
❑ Achieve continuous improvements in our environmental performance;
❑ Work in partnership with our professional agents, contractors, suppliers and tenants;
❑ Work with our clients to fulfil their environmental policy aims;
❑ Act honestly, openly and responsibly in our dealings with competitors, suppliers and customers;
❑ Influence others;
❑ Work in partnership with our professional agents, contractors, suppliers and tenants;
❑ Work with our clients to fulfil their environmental policy aims.

To implement this policy we will

❑ Identify key environmental and social objectives for each area of our business, and communicate with stakeholders;
❑ Set appropriate targets;
❑ Conduct regular reviews of our progress;
❑ Report publicly on our performance.

(4) Setting up systems to document your environmental programmes and procedures. All significant environmental aspects must be controlled effectively to reduce their impacts. A simple and effective way of providing this is to document best practice for communication for others to learn and follow from. A structured documentation system is paramount to simple on-going management – a common provision is to use a four-tier system comprising:

❑ A top level manual describing structures, roles, responsibilities and a vision for the system;
❑ A series of management procedures;
❑ Working instructions providing operational detail requirements; and
❑ Associated documentation including logs, records and supporting information.

It will be a major activity for facilities management to document the procedures and working instructions. It is important to define the scope of activities and responsibilities clearly within the procedures and to ensure an auditable trail is clear through the procedure from completed logs and records. The ISO 14001 standard does not require each step to be written unless it results in an environmental disaster. It is recommended including the key points and main aims of the activity, though leaving the procedure loose enough for minor changes to be made without having to change the procedure, reduces resources required to maintain the system. For many of the working instructions, these may be documented as part of the planned preventative maintenance (PPM) schedule.

(5) Setting objectives and targets to meet the significant aspects identified above. Where possible these should be kept to half a dozen at most and special care should be taken to ensure they are in keeping with the main business goals. It is likely that FM will be involved in the setting and achievement of at least one of these targets.

Typically objectives are set for areas such as waste, recycling, energy, hazardous materials, awareness and training. Targets are often set for reductions, however these reductions should be made against a known robust baseline. Therefore the first task with any of these is to develop the baseline.

(6) Establishing a programme to tackle your aims. An implementation programme is required detailing resources, skills and timescales to achieve the various elements of a system throughout the business. As part of this, specific off-site training may be required for one or more individuals within FM to manage the resulting activities in the system. There are a number of courses now available ranging from one-day overview sessions, through to five-day recognised audit courses and a plethora of diplomas and MSc courses to provide in-depth knowledge.

(7) Allocating roles and responsibilities – and providing sufficient resources. Once the programme has been determined, the development of the various committees with appropriate personnel is the next step to provide communication and commitment. A typical structure should have the majority of the roles provided between the steering and operational committees and the management representative. The roles and responsibilities should be communicated to all staff to allow local support to be provided where necessary.

(8) Ensuring all relevant staff are trained and adequately supported through the identification of training needs and the development of communication strategies to meet the requirements identified. A variety of training and awareness programmes are available to achieve this including CD-ROM presentations through the intranet, formal presentation sessions, booklets and posters or cascaded information through the management structure. The type of programme chosen should incorporate the management system set in place and what it is trying to achieve and the mechanisms to achieve this such as procedures and forms.

Providing training and awareness

The role of the training and awareness programme is about changing the culture of the organisation to move towards incorporating environmental issues into day-to-day practices and 'thinking green'. This is not a one-off activity but must be pursued on an on-going basis to ensure it remains fresh and interesting for staff.

A combination of programmes should be used to achieve this from:

❑ Development of an intranet page covering environmental activities and best practice, which is short, easily identifiable and encourages staff involvement;
❑ Poster campaign with a common thread attached to each poster such as characters or logos;
❑ Key point cards or booklets for staff to keep with them or refer to as required;

❏ Awareness presentations either formally or on screens describing environmental best practice applicable to staff; and

❏ Presentations through the intranet requiring completion of questions to verify understanding of the information.

(9) Setting up two-way mechanisms for communicating corporate responsibility issues internally and externally. These members will include those outside the business such as local communities, regulatory bodies and shareholders where necessary. Internal communication will involve ensuring progress is known and managing responses to queries. It is common for many of the staff questions to relate to FM issues such as waste management and energy conservation.

(10) Ensuring the actual documented operation meets your goals based on the procedures, objectives and targets and structures set in place. This should form the first part of an audit performed on the management system – either internally or through a third-party. Since the documentation will be the visible and promoted sign of the management system, it needs to be fully aligned and supportive of the objectives within the policy statement.

(11) Preparing emergency plans and documentation based on the significant aspects identified earlier. The plans must be tested for effectiveness on a regular basis. Emergency plans should include failures in gas and electricity supply, major legal non-compliances and major spillages, which all generally come under the remit of FM.

(12) Monitoring and measuring your organisation's performance, which will include legal requirements, benchmarking information and the objectives. Such areas will be total waste generated, waste recycled, water, gas and electricity usage, paper consumption, volatile organic compound emissions or total particulate emissions. This will be a time-consuming activity for FM, and is best performed in conjunction with general planned maintenance activities.

(13) Identifying and correcting failures and setting measures to prevent re-occurrence. A robust mechanism should be set in place to enable staff to identify failures within the system and report them to allow corrective and preventative actions to be taken. This enables trends to be identified to prevent re-occurrences at the core level.

(14) Keeping performance, legal and other records required to support the system as a whole. Records and their timescales should be noted within relevant procedures. It is recommended setting up some folders with dividers to collate and file all records in a central location. Where this is not possible, either from multiple sites, or from records within an engineering office, the location of all records and their keepers should be recorded.

(15) Periodic internal and external audits to measure the implementation of the system, on-going best practice, contractor and supplier management and legal compliance. The audits can be incorporated into existing reviews for workplace and engineering health and safety through the addition of additional sections.

(16) Management reviews to ensure continued improvements. This will be performed for the policy signatory to review trends from the audits and non-conformances and changes to the business which may affect the significant aspects, resources available and achievement of the objectives and targets set.

(17) Certification – companies can apply for accredited certification to the standard. However, not all companies see certification as the final goal. The development of the management system should be seen as the starting point from where improvements in terms of efficiencies, savings and product development can be derived.

2.4.9   Managing and continuous improvement

The ongoing management of a corporate responsibility management system will continue to require resources, time and effort in order to gain the benefits and efficiencies from the system. Commonly, there is a sigh of relief once certification has been achieved and the system left to operate itself until weeks, or even days before the surveillance visits. At this point, a large effort gets the system back on track in time for the audit. This process is followed every six months for the next three years causing a significant level of additional effort and providing little benefit overall for the business or the environment.

Organisations must have the resources available not only to implement the system, but to maintain it. For many organisations, the level required and resource usage over time is difficult to quantify both in the short and long term due to:

❑ Difficulty for companies to define resource initially;
❑ Companies unaware of resource usage over time; and
❑ Awareness over time becomes stagnant.

Many organisations require a major input of resources, both as money directly, but also in manpower to set up the system and ensure it is effectively operational. This commonly falls very rapidly, directly after certification has been achieved, with the misconception that the system will continue self-managing. Figure 2.12 shows the typical resource and effort provided which fluctuates widely during surveillance periods, when a panic peak occurs, and non-surveillance time periods.

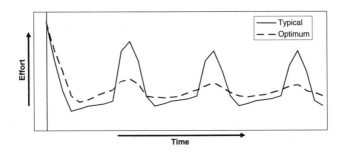

Figure 2.12   Typical and optimum effort post-certification.

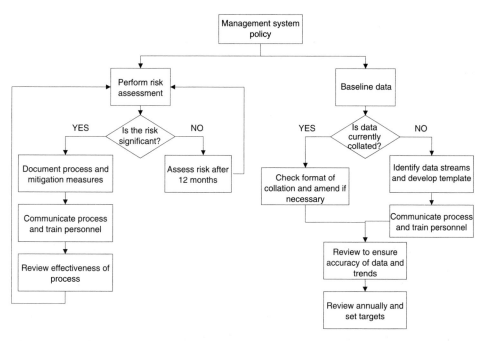

Figure 2.13 Continuous improvement in a management system.

It also describes the optimum resource profile where greater and more focused levels are provided throughout the time period resulting in less of a panic for the surveillance visit.

Whilst resources will remain higher due to the increased level of management and operational processes to ensure compliance to the management system, over time that increased resource throughout the year means less stress during the audit, and a higher chance of passing and achieving targets. Environmental awareness often falls because the emphasis is placed on performing the job, with the environmental issues captured as a bolt-on, and not crosses related back to the management system.

Following the development of a management system, the on-going management and compliance to the requirements will be audited on a regular basis based on the documentation and procedures provided. It will be the responsibility of the document owner to ensure they are kept up-to-date with any changes communicated through the document change process (Figure 2.13) and feedback through the system again.

Where possible, the activities required to manage the system should be scheduled into the planned maintenance regime on the timescales and intervals required. This will ensure that objective evidence is available, that all the activities have been performed, logs and records completed and documentation reviewed.

Additionally, awareness and knowledge of the system can stagnate over time, leading to insufficient resources provided to maintain the system and overlooking some of the benefits. Much of this can be mitigated through the development of the system in its initial form integrated into existing structures. Continual awareness through in-house magazines, achievements in reducing waste or energy in posters or the intranet are effective in maintaining knowledge of the system and the reasons why it has been developed.

Case study: Innovative approach secures ISO 14001 Certification – BAE Systems, Basildon

BAE Systems, avionics business at Basildon in the United Kingdom sought to achieve certification to ISO 14001 in the late 1990s. The Basildon site provides the infrastructure for a range of functions from the development of leading-edge electronics to office based activities. The design, manufacturing and repair centres house approximately 1500 staff at two locations – covering a total of 37 500m^2. At the time, the site was undergoing a major period of change as the organisation was transferred from part of the GEC-Marconi company to be merged with British Aerospace to form BAE Systems.

The initial brief was to perform an environmental assessment and to devise a strategy to achieve ISO 14001, anticipated to revolve around manufacturing activities on the site. A totally integrated approach was implemented, taking in the full product life cycle from inception, through design, to manufacture and ultimately disposal. This approach meant involving all the various business functions, suppliers and contractors including the facilities management (FM) and project management teams on site, to produce the most effective environmental management system.

There were a number of challenges to the implementation of the ISO 14001 system, possibly the greatest being the change in ownership within the company which led to difficulties in initially gaining support and then maintaining that commitment from staff, in particular, during the changes.

The full scope of activities on site included complex design and manufacture processes to meet client specifications with an increasing level of one-off products, so a standard procedure was not appropriate. The system also had to be capable of accommodating the rapid changes inherent in the businesses.

Development of the system

The starting point was to understand the current activities taking place on site, processes utilised and the environmental issues through a series of 1 : 1 meetings with key staff throughout the site. The interviews enabled an initial environmental review to be developed and provided the means for a debate within the business on the nature of the management system and programme to achieve certification.

A steering group, providing a core team of individuals, was developed to correspond to all parts of the avionics business at Basildon as representatives for the site management team. The steering group were involved in the development and testing of draft procedures. A series of amendments were made to ensure they were effective prior to cascading the processes effectively. The result was a system capable of reviewing every aspect of the company's activities to control and minimise their environmental impact.

Working in collaboration helped to generate innovative solutions. For example:

❏ A series of workshops was arranged involving experts from the manufacturing industry to look at new ways of working to construct and deconstruct products. As a result, environmental considerations were integrated into the design process which led onto reduced impacts during the manufacturing stages;
❏ Work with the on-site facilities management team helped to identify new ways to minimise pollution potential, and contribute to reduce costs. For example, rationalisation of waste contractors, improved handling of special wastes and more accurate forecasting of waste production; and
❏ A robust reporting and management structure was implemented to improve communication between the workforce and board, aiding continual improvement.

Communication

The core team also provided a means to communicate with staff at a local level within individual departments and relate directly to the environmental impacts of their roles and where benefits could be provided.

Culture change is a critical aspect to the implementation of an environmental management system proving a means for the processes to become bedded into day-to-day activities and part of operational life. The steering group provided the catalyst within their respective departments to enable this change in culture.

Alongside the steering group, the certification process created a core team within the environment, health and safety team who have a clear understanding of the environmental challenges facing the site and have the desire to drive continuous improvement – helping to minimise risk and costs.

Continuous improvement

Fundamental to the achievement of ISO 14001 is the evidence that continual improvement is being made. The site team delivered a number of successes not only through direct cost savings associated with waste and energy, but also indirectly through mitigating impacts as part of the operational functions on site.

The EMS is providing the infrastructure for continual improvement. New initiatives are being suggested at the staff level, such as a 'green transport programme', helping to improve the workplace and increase staff satisfaction. The design team at the site are setting new standards for the integration of environmental considerations, from first drawings right through to manufacture and beyond.

By rigorous monitoring of waste streams and increased focus on recycling opportunities, the business has made significant progress with its environmental management objectives. Latest figures reveal:

❏ The overall amount of waste produced has decreased by *41%* and waste transported to landfill has decreased by *33.5%;*
❏ Priorities for waste disposal are planned well in advance and better rates secured.

Finally, ISO 14001 gives BAE Systems avionics business at Basildon a compliant process when tendering for new work and assures the regulatory bodies that the site understands and complies with the latest environmental legislation.

2.5   Corporate responsibility reporting

Since the early 1990s there has been a dramatic increase in the number of environmental, social and corporate responsibility reports which are being generated by many of the world's largest companies. There is a feeling that the production of such documents is a pre-requisite to meet with shareholder, and other stakeholder requirements and gain a foothold within the 'ethical fund' lists. However, the take-up of these reports – those which are read and used – is limited, as is the information contained within these documents.

Facilities management and the property industry as a whole have a major role to play in the provision of accurate and transparent information.

Section learning guide

This section reviews the development of corporate responsibility reporting and the role facilities management can play in delivering these reports both internally and publicly:

❑ Background to corporate responsibility reporting including the drivers;
❑ Implications of the operating and financial review;
❑ Understanding the needs and requirements of stakeholders; and
❑ Developing a corporate responsibility report.

Key messages include.

❑ Reporting should be transparent – including the positive impacts made but also a recognition of the areas for improvement; and
❑ Reporting is for the stakeholder – define the stakeholder needs and provide the information in a succinct format.

2.5.1   Introduction

Corporate responsibility reports are defined as public reports by companies to provide internal and external stakeholders with a picture of corporate position and activities on economic, environmental and social dimensions. In short, such reports attempt to describe the company's contribution towards sustainable development.

Corporate reports covering non-financial information – variously called sustainability, triple bottom line (TBL), corporate responsibility (CR), and environmental, health, and safety (EHS) reports – are relatively young compared to financial reports. However, non-financial reporting is growing in significance as corporations and their shareowners and stakeholders recognise that non-financial issues impact financial performance.

There have been sporadic initiatives to produce non-financial corporate reports, such as the social reports produced in Germany during the 1970s. The current reporting movement emerged from reporting in the United States during the late 1980s, in response to the increasing volume of emissions data put into the public domain by the US 1987 SARA (Superfund Amendments and Reauthorization Act) Title III legislation – the 'right to know' legislation, which established the Toxic Releases Inventory.[28]

Also in 1987, the Brundtland Commission coined the term 'sustainable development'. Over the next several years, this concept created a ripple effect in non-financial reporting, spurring a transition from narrowly-focused reports to

---

[28] Non-financial corporate reports – http://www.enviroreporting.com/

ones that integrated diverse sustainability issues. During the early 1990s, and particularly following the 1992 'Earth Summit' in Rio de Janeiro, the existing environment and EHS reports began to include wider issues, such as community, and gradually 'sustainability reports' began to appear. The shift continued through the 'Rio + 10' summit on sustainable development in Johannesburg, South Africa in 2002. The percentage of reports exclusively focusing on the environment fell from 63% of non-financial reports in 2000 to 42% in 2002, while sustainability reports rose from 5% to 15% over the same period.

A 'one-size-fits-all' approach does not work for corporate responsibility reporting. Each company operates and performs differently with widely varying pressures and demands placed upon them. It is up to each company to determine the approach it wishes to take, depending on its situation and needs. Be it an environmental report, a social report, an environment, health and safety report or an integrated report – also called triple bottom line, corporate responsibility or sustainability report – all these various reporting formats contribute towards corporate responsibility reporting.

This is an evolving field. There are many companies that have yet to produce their first report while others have published environmental reports for many years and are now moving further down the road towards the more complex area of sustainable development reporting.

When non-financial reporting changed from being a short, glowing commentary on an organisation's philanthropic activities, embedded in the mid-pages of its annual report, into stand-alone sustainability reports, it was accompanied by a complementary development in assurance. Concerns about the impact of companies' production processes, labour standards in supply chains and human rights abuses in developing countries could not be assuaged solely organisational reporting. As in the financial sphere, information needed to be assured by an independent party.

External assurance became an essential part of the process, driven by the need for more credible information about performance both within and outside the organisation. Alongside the development of social and environmental and quality management audits, such as those of the ISO family, SA 8000, etc., emerged a way of verifying and improving the quality of an organisation's products and services. Assurance has subsequently been on the rise.

Having started off with a greater focus on 'accounting', or data quality, the interest in assurance has quickly shifted towards a more complex concept focusing on what really matters to stakeholders, i.e. stakeholder-based materiality, in order to handle the complex issues that sustainability raises.

While the value of accuracy focused assurance to ensure reliable and comparable data for management and information users still remains, stakeholders expectations have clearly shifted towards assurance processes that go beyond assessments of accuracy to exploring the quality of processes, such as stakeholder engagement, organisational learning and innovation as well as processes by which the organisation ensures strategic alignment with key stakeholder expectations (Figure 2.14).

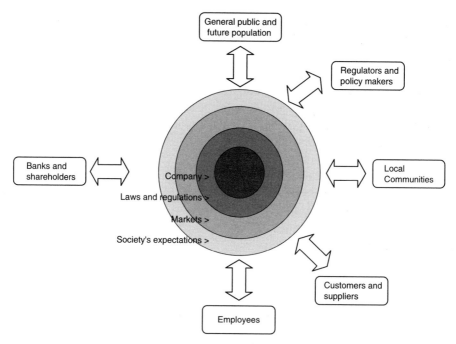

Figure 2.14   Corporate responsibility stakeholder needs and reporting development. (Adapted from *Beyond Reporting*, WBCSD, June 2005.)

In 2002, almost half of the Global Fortune 250 companies reported on their non-financial performance in some format, compared with 35% in 1999.[29] Research by UNEP and SustainAbility shows that of the top 50 companies globally, only 4% in 1994 had reports assured, which rose to 28% in 1997, 50% in 2000 and 68% in 2002.[30] Similarly, KPMGs international survey of sustainability reporting shows that in 2002, of the 112 GFT250 (Global Fortune) companies that issued a report, 33 (29%) had their report independently verified compared with 19% in 1999. In all, 25% of leading companies' reports was verified, with 65% of those being undertaken by the major accountancy firms.

The accountability rating, which rates sustainability reports from the Fortune Global 100 (or G-100, the world's 100 companies with the highest gross revenues), finds significant growth in sustainability reporting. Almost three-quarters (72%) of G-100 companies issued sustainability reports by 2004, whereas less than half (48%) had done so by 2003. The top ten sustainability report ratings included Hewlett-Packard, which was the sole US-based company, BP, Shell, Toyota and Unilever.[31]

Of the FTSE 250 companies, 54 reported in 2001, 132 reported in 2002 (100 social and ethical). Heavily polluting industry sectors including beverages, oil and gas,

---

[29] *International Survey on Corporate Sustainability Reporting*, KPMG 2002 report.

[30] *Trust Us* (2002), Global Reporters Research Programme by SustainAbility and UNEP – http://www.sustainability.com/

[31] AccountAbility rating – http://www.accountability.org.uk

personal care, tobacco and utilities provided full reporting, whilst sectors perceived to be lower polluting industries were poorer at reporting – IT, hotel companies and retailers. There has also been a split in the reporting with 80% of the top 50 European companies reporting, whilst only 22 of the top 50 US companies reporting.

The trend in western Europe and Japan is predominantly focused towards the three Cs – content, credibility, and communication, whereas the United States is more comfortable reporting how their money is spent, such as on charity and good works, rather than how is has been made.

The past few years have seen an increase in the number of companies who provide facilities management services reporting on environmental and social issues more fully. Johnson Controls released their first sustainability report in 2004, though Carillion were the first in the industry incorporating sustainability issues into their business model in the late 1990s.

2.5.2 Drivers

There are many reasons why companies choose to measure, manage and report on environmental issues, but recently increasing legislation and investor pressure have meant that more businesses have had to ensure that they are taking account of environmental issues. Leading companies are embracing open, transparent reporting as the foundation for their corporate (social) responsibility (CR) strategies as it sends a clear signal to the investment and wider community that addressing these issues is not merely an add-on policy but is integrated into their business activities. It is this approach that improves not only the company image but also its financial performance too.

(1) *Reputation*

❏ Differentiate your organisation and attract customers and suppliers;
❏ Enhance reputation as a responsible employer, improving staff retention, loyalty and recruitment;
❏ Changes to pensions funds and funds (FTSE4Good and Dow Jones Sustainability Index) is driving visibility and investment;
❏ It increases competitive advantage (the first mover effect);
❏ Public recognition for corporate accountability and responsibility;
❏ It may improve access to lists of 'preferred suppliers' of buyers with green procurement policies; and
❏ It enhances employee morale.

(2) *Communication and transparency*

❏ Convey organisation's values and promote two-way communication with stakeholders;
❏ Need to reveal where improvements can be made – transparency;
❏ Build trust with local and planning authorities, neighbours and community groups;

❏ Demonstrate to stakeholders of improving environmental performance;
❏ It demonstrates coherence of overall management strategy to important external stakeholders; and
❏ It strengthens stakeholder relations.

(3) *Risk management*

❏ Helps identify risks, and prevents damage to reputation from negative publicity on environmental issues;
❏ Protects brand;
❏ Effective self-regulation minimises risk of regulatory intervention; and
❏ It reduces corporate risk, which may reduce financing costs and broaden the range of investors.

(4) *Cost savings*

❏ Provide focus on environmental issues to identify efficiency savings;
❏ Target setting and external reporting drives continual environmental improvement; and
❏ Improved profitability.

2.5.3   Stakeholders and their information needs

Corporate responsibility reporting initially provided a single report for cater for the needs of all stakeholders, but primarily focused towards the shareholders and NGOs to demonstrate good practice. However, more recently there has been a need and requirement to provide information on an organisation's performance for a variety of stakeholders.

There are a wide variety of stakeholders who are interested in the environmental and social performance of facilities management activities – both in-house service provisions and outsourced service activities. The primary reason has been mentioned many times – the FM market employs some of the lowest paid staff, manages the environmental pollution aspects of most organisations and procures items and equipment from across the globe. Together, these impacts pose some of the biggest impacts on the corporate responsibility an organisation has.

A study among non-governmental organisations revealed that while 79% stated these reports are 'very' or 'fairly' useful, only 44% of the reports are considered 'believable'.[32] Using the internet and global media, NGOs have dramatically increased the pressure on companies to account for their performance. According to NGOs surveyed, the most important approach a company can take to improve a report's credibility is to acknowledge non-compliance, poor performance, or significant problems. This finding indicates that NGOs expect corporate responsibility communications to be straightforward and thorough. Other factors that boost

---

[32] Building CEO Capital, Burson-Marsteller – http://www.ceogo.com/documents/Building_CEO_Capital_2003.pdf

confidence are comprehensive performance metrics, third-party certification and standardisation of reporting. Less effective factors are case studies and independent research.

Consumers in North America, Europe, and Australia are consulting corporate social and environmental reports more commonly than originally thought.[33] According to the poll, half of the general public in North America, Australia, and some parts of Europe say they have either read a CR report themselves, briefly looked at one, or heard about one from someone else. Traditionally, reports are developed for investors and stakeholders primarily rather than the general public. Among those who are aware of CR reports, the majority say that reading or hearing about a report improved their impression of the company, led them to buy the company's products, or speak positively about the company to others. The publication of a CR report can impact corporate reputation and the bottom line, particularly if reports are tailored for and made available to the general public.

Typical stakeholders affecting FM include:

❑ Internal management to demonstrate accountability, compliance and progress to internal management processes;
❑ NGOs to build trust and greater two-way communication;
❑ Landlords, managing agents and other tenants to deliver good business practice as a collective rather than individual entities;
❑ Sub-tenants, suppliers and contractors to work with them to reduce the overall impacts;
❑ Neighbours/communities to work together and improve communication, reducing crime;
❑ Customers to understand needs and explain corporate responsibility progress achieved;
❑ Employees to explain and engender support for common goals; and
❑ Regulatory agencies.

An international assurance standard that advocates a reassessment of stakeholder-based materiality, AA1000AS, was launched in 2003. It enabled an auditor to assess an organisation's performance against sustainable development commitments, policies, and strategies as well as stakeholder expectations and behaviour using three principles core to assurance.[34] The methodology provides generally applicable principles for assessing, attesting to, and strengthening the credibility and quality of organisations' sustainability reporting and associated management processes, systems, and competencies. As such it can be operated in line with the global reporting initiative to both develop and check CR reporting.

The three principles of materiality, completeness, and responsiveness provide the framework within which an auditor can assess the report and related sustainability

---

[33] *Building a Dialogue with Stakeholders on CSR Reporting*, GlobeScan Inc – http://www.globescan.com/csrr_analyzer.htm

[34] AccountAbility AA1000AS – http://www.accountability.org.uk

management processes. Materiality requires indicating how one defines which stakeholder issues are material to the business, for example, compliance with law, industry codes, internal policy, or stakeholder expectations. These elements may vary across organisations' end-markets according to regulatory, cultural, or geographical contexts. Completeness involves indicating how a company understands and measures performance, impacts, and stakeholder views relating to identified material issues. This includes performance relating to organisational activities, products, services, sites, subsidiaries, etc. Responsiveness requires indicating how one responds to material stakeholder concerns and interests in a timely fashion, and how sufficient resources are allocated to implement related policies, commitments, etc.

2.5.4   What to report on

Key to successful reporting is the notion of stakeholder engagement and its role in both sustainability management and related assurance processes. Stakeholder engagement should be at the core of any report to firstly provide tailored information that is specific to diverse organisations, their business sectors, stakeholders and business strategies and models, and secondly because good-quality stakeholder engagement processes can powerfully inform internal decision-making and enable learning and innovation.

Ultimately reporting is based on perception – both that of the company reporting and the receiving stakeholder. The value of engagement is to provide a more consistent perception.

Typical information to include in a CR report is provided in Table 2.6 which captures the social, economic and environmental aspects against the 14 sustainability categories used throughout this book. It should be noted that the table is not exhaustive and should not be taken as such. For each of these categories, discussions with the various stakeholders on the most relevant information should take place.

Mandatory corporate reporting on social and environmental issues is stalled in limbo in the United States, but it is proceeding forward in the United Kingdom through changes to Company Law requiring greater information on environmental and social risks to be tangibly provided in the annual report.

❑ The purpose of reporting is to help shareholders assess a company's strategies and the potential for them to succeed;
❑ Requires directors to provide shareholders with a balanced and comprehensive analysis of their business and future prospects;
❑ To include information about the environment, employees and social and community issues – or make a positive statement if not disclosed; and
❑ Company directors carry personal liability and potentially face unlimited fines for non-compliance.

There is a drive for directors to provide and analyse both the current and future trends and factors underlying performance of the business. The intention

Table 2.6 Typical social, economic and environmental criteria to report on

Sustainability category	Social aspect	Economic aspect	Environmental aspect
Management	Ranking as an employer	Net profit, EBIT, dividends Tax paid per country	Compliance
Emissions to air			Carbon dioxide emissions – buildings and fuel Ozone depleting substances used
Land contamination		Remediation costs	Land use and potential contamination
Workforce occupants	Job satisfaction Retention of staff Health and safety incidents Training received Grievances upheld	Labour productivity	
Local environment and community	Community engagement	Jobs by country Philanthropic/ charitable donations Local suppliers used	Complaints upheld by local community
Life cycle or building/products		Future liability costs – contamination, flooding Effects of climate change	Recycled materials purchased Hazardous materials purchased
Energy management		Cost of energy	Energy consumed Renewable energy purchased
Emissions to water		Cost of water	Impact to waste sources Discharges to water
Use of resources		Cost of materials	Total materials used Water consumption

Continued

Table 2.6 *Continued*

Sustainability category	Social aspect	Economic aspect	Environmental aspect
Waste management		Cost of waste disposal	Volume of waste generated and destination Volume of waste re-used
Marketplace	Performance of suppliers	Performance of suppliers	Performance of suppliers Impact of products
Human rights	% below minimum wage Ratio between highest and lowest wage Number of human rights violations	Total wages by country	
Biodiversity	Access to biodiversity		Impact on biodiversity from occupation of land
Transport			Green travel programme

is to encourage companies to be increasingly open and transparent, which helps investors to assess the strategies adopted by the business and the potential for these strategies to succeed.

2.5.5   Guidance on reporting process

There is a four-step process to sustainability reporting. First, corporations produce sustainability reports using whatever information can be gathered. Second, they develop systems to gather the relevant information more comprehensively. Third, they audit the information-gathering systems and the data internally. A third-party audit is the fourth step. Often, stakeholder engagement is the prequel, or occurs before the first report. Thereafter they set up stakeholder panels or focus groups to maintain the process of dialogue and feedback.

Obstacles identified by the report include lack of consistency in approaches to external assurance, which hampers credibility; weak support from governments guiding or mandating sustainability reporting; and slow buy-in by mainstream investors. Exemplary sustainability reports address issues such as reporting processes, assurance mechanisms, and the business case for sustainability reporting. For example, the report cites a statistic from the Anglo American Platinum of South Africa sustainability report that the company's HIV/AIDS programme for employees saved the company 42 million Rand in 2003 and may save 1400 lives a year.

The global reporting initiative (GRI) has documented guidelines on how to develop CR reports.

(1) Identify your audience to focus efforts, tailor the report format and provide applicable information;
(2) Review the main environmental, social and economic impacts and write a CR policy if applicable;
(3) Consider what to include in the report, based on the amount of information available and progress and reliability of data collection. An incremental approach may be desirable to increase scope, data collection and impacts, though this must be explained alongside timescales, priorities and progress;
(4) Choose the method of reporting, e.g. included in annual report, stand alone report or a web based version. This will be dependent on existing mechanisms, the audience, and resources; and
(5) Provide assurance for reports through third-party verification based on how other sector based organisations provide theirs, internal systems in place, what stakeholders expect, resources and timescales.

Environmental report – contents

(1) Statement from MD/CEO including an endorsement of the policy, highlights, successes and failures, plans to improve and vision/mission for managing issues in the future;
(2) Profile of organisation:

❑ Name and scope of activities;
❑ Contact details;
❑ Establishment and funding details;
❑ Functions and business operations (products and services);
❑ Size, sector, markets, customers, geographic scope;
❑ Details of reporting period, significant changes since last report (normalise performance measures); and
❑ Detailed explanation of which operations the report covers, and any exclusions.

(3) Executive summary and key indicators in tabular/graphical format;
(4) Vision and strategy – sustainability and eco-efficiency;
(5) Copy of the CR policy, description of the management system, including briefly roles and responsibilities, committees, management and monitoring of policy requirements;
(6) Key environmental, economic and social impacts (keep to between 6 and 8) – including stakeholder relations and communication;
(7) CR performance indicators based upon the impacts identified above as the basis to measure current status and improvements. These indicators should

be recognisable to the stakeholder, e.g. number of health and safety incidents, carbon dioxide emissions, employee satisfaction;
(8) Targets/improvements;
(9) Legal compliance issues and how they are managed and maintained – include any previous prosecutions;
(10) Independent verification statement.

An electronic reporting network has been developed to support and capture social, environmental, economic and corporate governance information for investors, the financial community and other stakeholders. Over 50 multi-national companies from around the world, including 22 Fortune 100 companies, participated in the development of OneReport^{TM}.[35] The network connects publicly traded companies with the firms that research corporate social and environmental performance for institutional investors, holding details used for many of the corporate responsibility indices. OneReport can also be used to create reports based on the global reporting initiative (GRI) guidelines, helping to understand the scope of GRI reporting and identify data that can be included in their GRI report. AccountAbility's AA1000 assurance standard provides the basis for additional functionality that will assist users in determining the extent and quality of assurance of the data.

The outputs from the report generated will need to meet up with the following requirements:

(1) *Quantification*: The significant impacts identified need to be defined fully and expressed in terms to enable benchmarking;
(2) *Coverage*: Reporting needs to cover all operations in which the company has an equity stake for each period of reporting, including both direct and indirect emissions, and, where relevant, estimating emissions from products and services;
(3) *Comparability*: Common metrics are needed to enable comparisons such as those made in this report, between companies, and between companies in different sectors;
(4) *Timeliness*: Disclosure should also be published at the same time as financial reports and relate to the same time period; and
(5) *Targets*: Clearer guidance is also required on the objectives companies are setting themselves for meeting their targets from both products and processes in ways that can be measured over time.

2.5.6   Different approaches to reporting

There is a variety of reporting formats which can be provided. At one extreme is a report detailing compliance to legislation and at the other is an open CR report.

---

[35] Electronic Reporting Network for social, environmental, economic and corporate governance information – http://www.one-report.com

Increasingly reports are produced through the internet, allowing customised versions to be made available for the various stakeholders rather than the historic one-size-fits-all printed document.

❏ Compliance – based upon legislative requirements, e.g. water and utility providers;
❏ Toxic release inventory – based upon emissions of toxic substances;
❏ Eco-balance – based upon mass balances of inputs and outputs from which performance indicators are derived;
❏ Performance – structured around the major and significant impacts on the environment: performance measures are set and reported upon;
❏ Product focussed – review the total environmental impacts of products including operation, recycling, manufacturing and environmental management;
❏ Environmental and social reporting – including employee statistics and conditions, community support, and stakeholder consultation information;
❏ Corporate responsibility reporting – full disclosure of corporate practices and programmes taking into consideration future liabilities.

Case study: Sustainability Report 2004[36] – Gap Inc.

On 12 May 2004, Gap Inc. released a report that had jaws dropping in corporate boardrooms and activist corridors across the land. The 40-page 'social-responsibility report' details, with unflinching honesty, the problems the $6.5 billion clothing retailer found in the 3000 factories it contracted to produce clothing for its Gap, Old Navy, and Banana Republic brands.

The company discovered persistent wage, health, and safety violations in most regions where it does business, including China, Africa, India, and Central and South America. The infractions range from failure to provide proper protective equipment to physical abuse and 'psychological coercion'. Though discoveries of the worst violations (such as child labour) were rare, Gap reported that it had pulled its business from 136 factories and turned down bids from more than 100 others when they failed to meet its labour standards.

None of the findings were especially surprising; labour abuses are a fact of life in the global apparel industry, where intense price competition continually drives factories to produce more clothing for less money. What was extraordinary was Gap's willingness to go public and reveal, in exceptional detail, its responses to these conditions. Even some of Gap's harshest critics say the company's candour will drive industry changes that ultimately improve the lives of factory workers.

Often, courage evolves over time; at Gap, courage is very much a work in progress. Gap published the report only after Domini Social Investments and

---

[36] GAP Inc. Corporate Sustainability Report – http://www.gap.com

other investors filed a shareholder resolution requesting greater transparency from the company. But in rising to meet that demand, Gap offered not the sanitised gloss that might have been, but a warts-and-all profile that freely admits to not having all the answers.

Its first response to the sweatshop exposés of the mid-1990s was to clam up and go into fix-it mode. It built an elaborate monitoring system, which today has more than 90 members who perform more than 8500 factory inspections each year. Gap has gradually realised that internal monitoring alone cannot unravel the industry's tangled problems, but that increased openness is just a first step. Gap must go further if it expects to build on the goodwill the report has engendered. And, impressively, it has committed to do so. The report outlines next year's goals, including allowing an external review of its own monitoring. Most significantly, Gap has agreed to rethink accepted garment-industry business practices, which include unrealistic production cycles that drive such abuses as unpaid overtime.

3 Facilities life cycle

This chapter will review the inclusion of sustainability criteria into capital projects, from the acquisition, design and construction or refurbishment of properties. During these activities FM will have a role to support the initial brief definition, and how the various measures are implemented to reduce operating costs and deliver a more sustainable facility.

Different industries such as building and construction, railways, highways, water, aerospace, manufacturing, oil and gas have been applying varying asset management principles to their operations for years to increase profitability, manage activities better, achieve efficiency gains and safety performance targets. In the local government arena, the need to demonstrate value-for-money and best value in procurement of services is a major driver to have asset management plans in place. Moreover, the increased implementation of Design-Build-Finance-Operate (DBFO) procurement methods forces long-term thinking, as consortia are required to operate assets typically for 25–30 years.

On the other hand, infrastructure owners are struggling with challenges such as escalating maintenance costs and lack of reliable knowledge of asset condition and performance at a given time. Different industries will have varying infrastructure management principles and processes in different contexts. Effective and efficient infrastructure asset management and whole-life costing are becoming increasingly important in any owner's investment decisions. Asset owners are therefore recognising the benefits of managing assets along whole-life principles. There is a need for a holistic approach that can bring significant value, and more effective infrastructure asset management in the long term under the following key areas:

❏ Framing the requirements – qualitative feasibility study to define needs and outcomes;
❏ Scoping the decision – often made in a project definition study looking at a cradle-to-cradle approach;
❏ Planning and design – analytical decisions made in the detailed design stage to review options;
❏ Implementation, delivery and operations; and
❏ End of usable life.

3.1 Life cycle facilities approach

The delivery of sustainability within a facility is best achieved through the entire life cycle to provide maximum returns. There is a widely acknowledged diminishing return and increasing cost through a facilities life cycle to retro-fit sustainability. Whilst this notion is accepted, it is surprising how late environmental and social aspects are discussed with respect to the operation of the facility. Within each of the phases of a facility, there is the opportunity to encompass sustainability criteria to deliver an improved asset for both staff operating within and book value.

An international study, led by the Royal Institution of Chartered Surveyors (RICS), found that green buildings can:[1]

❑ earn higher rents and prices;
❑ attract tenants and buyers more quickly;
❑ cut tenant turnover;
❑ cost less to operate and maintain; and
❑ benefit occupiers.

This study shows that the interests of business and the environment can converge – achieving real environmental benefits can also be profitable. The study concludes that 'more work needs to be done on the value of green buildings, but that the findings are encouraging: Evidence that sustainable practices can add value supports the claims and direction of the green building industry. This is an important step towards greater acceptance of green buildings in the marketplace'.

Section learning guide

This section provides an overview to the building life cycle and the linkages between each of the various stages. This takes into account the initial stages in briefing and feasibility studies, through the traditional design and construction activities and onto operation and capital project delivery:

❑ Introduction to a sustainable life cycle approach;
❑ Overview to the history of green buildings;
❑ Sustainability involvement throughout the life cycle;
❑ Describe the benefits of early intervention; and
❑ Whole-life value concept.

Key messages include:

❑ Early involvement of FM will provide a better operational facility;
❑ Challenge the traditional concepts to designing buildings; and
❑ Include the community and staff in the decision-making process.

---

[1] Royal Institution of Chartered Surveyors Green Value Report – http://www.rics.org/greenvalue

3.1.1   Sustainable buildings

What is a green building? The US Office of the Federal Environmental Executive defines green buildings as 'the practice of (1) increasing the efficiency with which buildings and their sites use energy, water, and materials, and (2) reducing building impacts on human health and the environment, through better siting, design, construction, operation, maintenance, and removal – the complete building life cycle'.[2]

The gap towards a more sustainable building is the involvement of the supply chain, stakeholders and the impact on the local community such as the economy, skills and working practices.

The development of the modern day building – an air-conditioned sealed box with fluorescent lighting and ventilation systems has only been in place since the 1930s. Prior to this, facilities used passive systems to provide ventilation, deep-set windows to shade the sun and windows that opened. As far back as the nineteenth century, structures like London's Crystal Palace and Milan's Galleria Vittorio Emanuele II used passive systems, such as roof ventilators and underground air-cooling chambers, to moderate indoor air temperature.

From the late 1960s thoughts of alternatives in continuing to build in a sealed box manner were inspired by the work of Victor Olgyay (design with climate), Ralph Knowles (form and stability), and Rachel Carson (*Silent Spring*). During the mid-1970s, characterised by a global energy crisis, innovative designs were taking hold. In England, Norman Foster used a grass roof, a daylight lit atrium, and mirrored windows in the Willis Faber and Dumas Headquarters (1977). California commissioned eight energy-sensitive state office buildings, notably the Gregory Bateson Building (1978), which employed photovoltaics, under-floor rock-store cooling systems, and area climate-control mechanisms. At the international level, Germany's Thomas Herzog, Malaysia's Kenneth Yeang, and England's Norman Foster and Richard Rogers were experimenting with prefabricated energy-efficient wall systems, water-reclamation systems, and modular construction units that reduced construction waste. Scandinavian governments set minimums for access to daylight and operable windows in workspaces.[3]

On 'Earth Day' in 1993, US President Clinton announced plans to make the presidential mansion 'a model for efficiency and waste reduction'. The 'greening of the White House' involved an energy and environmental audit, and a series of design charettes in which nearly a hundred environmentalists, design professionals, engineers, and government officials were asked to devise energy-conservation solutions using off-the-shelf technologies. Within three years, the numerous improvements to the nearly 200-year-old residence led to $300 000 in annual energy and water

---

[2] *The Federal Commitment to Green Building: Experiences and Expectations*, 18 September 2003 – http://www.ofee.gov/sb/fgb_report.pdf

[3] EU Design Practices – http://www.unep.or.jp/ietc/sbc/index.asp

savings, landscaping and solid-waste reduced costs, while reducing atmospheric emissions from the White House by 845 tons of carbon a year.[4]

The success of the 'greening of the White House' programme promoted the greening of other US federal buildings including the Pentagon and National Parks including Yellowstone. This work was consolidated in 'Greening Federal Facilities',[5] an extensive guide for federal facility managers, designers, planners, and contractors, produced by the Department of Energy's Federal Energy Management Programme.

In the first instance, property clients and investors need to consider whether they can re-use or recycle existing buildings through refurbishment or conversion to other use. The potential for the development of 'universal buildings' where their design can be varied to take account of the changing urban environment and activity needs to be more thoroughly considered. This is common across the world where many cities that date back a few hundred years have premises that have changed use over time. Many of the premises, such as eighteenth-century properties in London, were originally designed for residential use and subsequently changed to retail and now many provide hotel accommodation.

Over the past 50 years across Europe a number of best practice techniques have been tried and implemented as standard principles e.g. modular construction processes in Germany, district heating and cooling systems as standard in Sweden, air tightness measures in Norway and passive cooling technologies across Mediterranean countries.[6]

3.1.2   Building life cycle

The building life cycle from a facilities manager's perspective does not begin at building handover, but at the initial briefing stage where decisions on funding, operability and life cycles are determined. The value of incorporating the end-user perspective at this stage provides intimate knowledge of the organisation and culture. Liaising with the rest of the project team, this knowledge will help to challenge the pre-concepts often put forward at the initial stages of the design.

Sustainability considerations within the life cycle of a building are described in Figure 3.1 which shows the sustainable construction potential against the facilities life cycle; this can be modified and improved at various key meetings and phases throughout time.

Initially, the sustainable construction potential (SCP) is high at the start of the project with the greatest level of flexibility and ease of changing decisions without significant cost implications. However, as the project progresses, the SCP decreases over time. The level of this decline will be variable and is dependent upon decisions made during the development of the design, and changes in construction process.

---

[4] US Federal Energy Programme – http://www.eere.energy.gov/femp/pdfs/greening_whitehouse.pdf

[5] US Greening Federal Facilities Guide – http://www.eere.energy.gov/femp/

[6] Modular Design Construction – http://www.connet.org/uk/bp.jsp

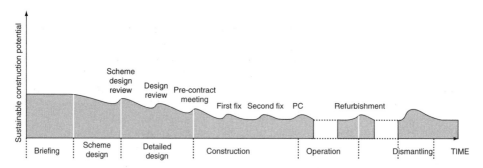

Figure 3.1   Sustainable construction potential through the building life cycle. (Source: Trevor Butler, Building Design Partnership.)

Throughout time, the SCP peaks during reviews and meetings, as sustainability measures are incorporated back into the project. The most striking factor is the change which can be generated at the briefing and feasibility stages by the inclusion of sustainability practices.

Table 3.1 below represents the various opportunities available to influence sustainability throughout the life cycle of a facility. Sections 3.2–3.5 cover each of these opportunities in more detail. Importantly, the decisions made at the briefing stage will affect the later design, construction and operation of the facility.

The functionality of the facilities manager in conjunction with the project team is described in Figure 3.2. The matrix covers the social, environmental and economic aspects throughout the life cycle of a facility with each of the aspects split into a series of sub-sets to define the role of the wider built environment. The various functions involved are included at the top from the investment phase through to demolition. Each function will have a varying role to play, represented by the asterisk – the greater the number of asterisk's the greater the role. For the FM function, involvement is required in all aspects, with a focus on the environmental and economic activities.

3.1.3   Whole-life value

The life cycle of facilities has traditionally been separated into the initial construction investment and procurement capital outlay followed by the revenue stream comprising day-to-day operational management and small capital investment. The gap causes pressure on the ownership of sustainability practices with regard to who will pay for them and how they will be carried out. Traditionally higher upfront costs will provide no benefits for strict capital investment budgets in speculative developments.

In recent years, there has been a move towards a whole-life value approach in public and private sector investment decisions. Many of the drivers have been discussed earlier (Section 2.3) and these include cost, transparency, stakeholder and risk management factors. A recent UK National Audit Office report (2005) states that design, procurement and decision-making need to be based on whole-life value

Table 3.1 Sustainability opportunities throughout the facilities life cycle (Source: Sustainable Construction Procurement CIRIA Publication C571)

Key project stages	Opportunities for influencing sustainability issues in a project
Define needs / briefing	Sustainability objectives
	Funding availability – ring-fenced monies
	Co-ordination with corporate responsibility agenda
Feasibility studies	Sustainability impact appraisal of alternatives:
	❏ routes, sites, technologies
	❏ new-build versus re-use
	❏ demolition
	Life-cycle cost studies
	Selection of advisers and design team
Decision to construct	Brief writing, including sustainability goals, targets, etc.
	Stakeholder engagement
Outline designs	Project sustainability policies
	Innovative design solutions
	Performance specifications
Planning permission	Environmental impact assessment
	Sustainability appraisal
	Public/community engagement
Scheme design stage	Performance specification for systems and products
	Life-cycle and cost analysis
	Value engineering
Construction tendering	Materials/component specification
	Contractor selection based on sustainability issues
Construction	Construction planning
	Sustainability/environmental management plan
	Waste/material management
Handover and commissioning	Final sustainability performance assessment/verification
	Energy/building management system
Fit-out	Procurement of furniture and materials
Occupation	Performance in use
	Post-occupancy evaluation
	Capital projects
	Operational management
	Churn
Decision to refurbish	Sustainability evaluation of options
	Adaptation for new use
	Demolition for recycling
Demolition	Re-use and recycling of materials

COMPONENTS OF SUSTAINABILITY	ABILITIES	Investment Function	Insurance Function	Planning Function	Client Function	Procurement Function	Design Function	Specification/Preliminaries Function	Construction Function	Regulatory Function	Supply Function	Operations & Management Function	Demolition Function
To achieve Sustainability by: The 'Functions' need to be able to:													
SOCIAL													
Optimising opportunities and social benefits	create usable public and private space to deliver successful communities	**		***	***	*	***	*	*	***	*	*	**
	improve health, wellbeing, accessibility and security of community	**	*	**	***	*	**	*	**	***	*	*	**
	enhance employment and skills development opportunities for the local community	**		**	***	**	**	**	**	*	**	***	*
Promoting sustainable communities through planning and design	meet requirements of local regional and national development and regeneration strategies	**	*	***	***		***	**		***	*	**	
	ensure appropriateness of development to needs of the community including multiple use and adaptability	**		***	***	*	**	*		***		**	
Engaging Stakeholders	consult with the public authorities, general public and other stakeholders, including end users and respond accordingly	**	**	***	***		***	*	***	***		**	**
	involve and manage expectations of stakeholders in development process from concept to commissioning	**	**	***	***	**	***	*	**	**	**	***	*
	consult and manage expectations of stakeholders on changes to ongoing use and operation	*			**	**	**	***		**	***	***	
Minimising negative impacts	plan for effective public and private transport use		*	***	*	*	**	**	***	***	***	**	***
	control nuisance (noise, dust, light etc)		*	***			**	***	***	***	***	*	***
	ensure a secure side, in construction			*		**	*	**	***	***	***		***
	ensure health and safety of site workers and local community	*	*				***	***	***	***	**	*	***
	protect, enhance and maintain appropriate social access to environmentally sensitive areas	**	**	***	**		***	***	***	***	*	**	***
	assess and mitigate flood risk	**	***	**	*	*	***	***	**	***	*	**	*
ENVIRONMENT													
Taking account of natural capacity	assess and mitigate wider environmental impacts (e.g. water supply, sewerage, transport, waste, etc.)	**	*	***	**		***	*		***	***	*	
	respond to projected impacts of climate change	**	***	***	**	**	***	**	*	**	***	*	
Optimising environmental benefits	minimise energy demand and meet it efficiently aiming to achieve carbon neutrality	**		**	***	**	***	***	***	***	***	***	***
	minimise water demand and aim to maintain water sufficiency from public supply	**		**	***	**	***	***	***	***	***	***	***
	optimise efficiency of materials use	**		**	***	**	***	***	***	***	***	***	***
	maximise range of environmental benefits in the	**	**	**	**	*	***	***		***	***	**	
	maintain and enhance biodiversity			**	**	*	***	***	***	***	**	**	*
Minimising negative impacts	reduce, reuse, recycle, recover waste	*		**	**	**	***	***	***	***	***	***	***
	reduce emissions to air, land and water	*	*	*	**	**	***	***	***	***	***	***	***
	reduce transport impacts			***	**	**	**	***	***	***	***	***	***
	protect ecological resources			***	**	*	***	***	***	***	***	***	***
	minimise take of environmentally valuable land	*		***	**		***			**	***	*	*
	minimise pollution of air, land and water	*	**	*	**		***	***	***	***	***	**	**
	manage and control in situ contamination of land	**	**	***	**	*	***	***	**	***		**	**
	protect archaeological and historically valuable resources	**	**	***	**	*	**	***	**	***		**	**
ECONOMIC													
Ensuring economic viability and improving processes	use technologies and materials consistent with sustainability principles	**	**	*	**	***	***	*	***		***	***	***
	keep up-to-date with advances in construction/technology	**	**	***	**	***	***	***	***	**	***	***	***
	establish cost & benefit on whole life value basis	**	**	***	**	***	***	***	***	**	**	**	**
	manage the supply chain effectively					***	*	*	***	**	***	***	**
	keep up-to-date with regulatory and planning requirements	*	**	***	**	***	***	*	***	***	***	***	***
	operate effective project management and contingency planning procedures		**	**		***	**	*	***	**	***	**	***
	maximise range of economic benefits including flexibility of use	**	**	***	**	**	***	*		***	**		
	achieve cost effective out-performance of statutory requirements	*	**	**	**	***	***	*	***	***	***	***	**
Enhancing business opportunities	meet requirements of national, regional and local economic strategy	*	**	***	**		**	***	*	***	*	*	
	capitalise on funding/grant available for more sustainable development	**	*	*	**	***	*	*	**	**	**	**	**

Figure 3.2   Role and inter-relationship of the facilities manager as part of the whole building process. (Source: Skills Sub-Group, UK Sustainability Forum.)

covering more than just the costs associated with the acquisition and operation of an asset.[7]

Whole-life value (WLV) encompasses economic, social and environmental aspects associated with the design, construction, operation, decommissioning, and, where appropriate, the re-use of the asset or its constituent materials at the end of its useful life. WLV takes into account the costs and benefits associated with the different stages of the whole life of the asset. The WLV of an asset therefore represents the optimum balance of stakeholders' aspirations, needs and requirements, and the costs over the life of the asset.[8] There will be trade-offs between the various short-term project constraints (such as time, costs and quality) and the conflicts in stakeholders' longer-term interests and objectives.

WLV is of more importance than value for money (VFM) for one simple reason – WLV represents the long-term value for the money invested whereas VFM tends to represent the immediate spend. As such, sustainable FM considerations are better developed using the WLV methodology.

Design, build, finance, operate (DBFO)

In DBFO procurement, the number of projects has increased steadily in different industry sectors in a model theoretically expected to deliver greater WLV than traditional procurement methods, due to expected efficiency gains and reduction in costs resulting from the sharing of knowledge and skills in design, construction and operation. There is increased focus on service delivery over 25–30 years to a defined standard based on output specifications and not just delivery of the asset or project. Improvement in efficiency is expected by exploiting the private sector's managerial practices, ability to innovate and to take risks. The private sector service provider is expected to provide innovative methods of delivering the service, thereby reducing whole-life costs. A clear understanding of risk in conjunction with the enhancement of long-term value is essential for achieving WLV.

A study by the Centre for Policy Studies for UK Private Finance Initiative projects[9] identified that 88% had been delivered on time or early and with no cost overruns. This compared with 30% public sector projects delivered on time and 73% completed over budget. The implications are that better value is provided through private finance initiatives.

In the private sector, the drive is to get owners and clients to measure projects on value rather than cost alone. Aspects that are measured include:

❏ Procurement to be determined on the basis of quality, likely cost-in-use, out-turn price and known past performance as well as price;

---

[7] *Improving Public Services Through Better Construction*, UK National audit office – http://www.nao.org.uk/publications/nao_reports/04-05/0405364.pdf

[8] Bourke, K., Ramdas, V., Singh, S., Green, A., Crudgington, A. and Mootanah, D. (2005), *Achieving Whole Life Value in Infrastructure and Buildings*, Building Research Establishment, Garston, Watford – http://www.brebookshop.com

[9] Craig, A. and Roe, P. *Reforming the Private Finance Initiative*, Centre for Policy Studies – http://www.cps.org.uk/

❏ Construction should be designed and costed as a total package, including costs in use and final decommissioning;
❏ WLV to be appraised and the supply chain to commit itself to build on time, to budget and quality, and provide genuine value for money throughout the life of the construction; and
❏ Advice (to clients) should cover a range of procurement and management options, including environmental performance, operating and WLV.

Critically, the DBFO has raised the need for a strong personal relationship between the client and service provider, whether from regular meetings or through being colocated in the same space. Despite the growing body of evidence, some clients are unwilling to progress further than the initial thinking. Projects commonly require a high level of professional fees to provide the front-end risk evaluation and mitigation. Uncertainty and necessary innovation adds to the costs in both the initial fees and throughout the project. Interest rates need to be factored in and are commonly higher than the public sector would set.[10]

As with any prediction of the future, WLV analysis is a guess – an educated one most certainly – but a guess none the less. The analysis calculations depend on the prediction of future interest rates and future costs, so it is easy to see why there is little certainty in the outputs and why those outputs cannot be used for exact budget formulation. The most appropriate purpose of WLV is economic comparison. The variations in interest rates are minimised as the same variations will apply across all alternatives being considered. The choice of the discount rate (interest rate) used can have a dramatic effect on the outcome of the analysis. As an example, an annual energy bill of £100 000 over 30 years will have a present value of around £1.7 million if a 3.5% interest rate is taken, but only £600 000 at 1.5%.

Whole-life value incorporates factors that drive value associated with the commissioning and operation of an asset, including the values held by the different stakeholders. In a typical 30-year project, the value of operating the facility can be the same or greater than the initial capital outlay. The inclusion of sustainability criteria at the front end can deliver significant savings over the lifetime of the facility.

There are three main criterion used in determining whole-life value to select the preferred solution. The final choice of scheme to be implemented will be a compromise between these three:

❏ The lowest whole-life cost;
❏ Technical evaluation; and
❏ Environmental and social evaluation.

It is important that the facility manager be involved in the data acquisition process early in the design phase and that the relevant schedules and costs be estimated

---

[10] *Effectiveness of Operational Contracts in PFI*, Business Services Association and KPMG, 2005 – http://www.kpmg.co.uk/email/iandg/pfi_report.pdf

as accurately as possible. There are a number of errors usually made by practitioners when carrying out a whole-life cost analysis which can be easily rectified:

❏ Omission of data;
❏ Lack of a systematic structure or analysis;
❏ Misinterpretation of data;
❏ Wrong or misused estimating techniques;
❏ A concentration of wrong or insignificant facts;
❏ Failure to assess uncertainty;
❏ Failure to check work;
❏ Estimating the wrong items; and
❏ Using incorrect or inconsistent escalation data.

Case study: Corporate sustainability strategy – Carillion

Few organisations have incorporated sustainability into business operations as fully as Carillion has. Whilst being transparent and open about progress, admitting the path ahead is a long one, the approach being taken is seen as a signpost for the FM industry to follow. Carillion is a global construction and facilities management organisation.

The 'light bulb' moment came in 1994 at Twyford Down during the protests on the new road scheme through the area. Carillion, known as Tarmac at the time, were particularly in the public eye and castigated heavily for the damage on the environment and society that they were causing. The extreme public pressure and business impacts provided a profound wake up call to the Tarmac management team.

In the following five years, Tarmac took up environmental issues at a strategic level not only to repair the company's reputation, but as a core strategy of the business model, culminating in a series of awards and becoming sector leader towards the end of the century.

Corporate strategy

In 2000, reflecting the growing understanding of sustainability in industry, Carillion began to look at the ways in which it could become a more sustainable company. These are the steps Carillion has taken:

Step 1: Sustainability policy

To provide a focus to a suite of related policies (environment, human resources, health and safety) Carillion developed a sustainability policy to bring together the vision, values and objectives of the business under the guiding principle of 'sustainable solutions for the way we live'.

LIVERPOOL JOHN MOORES UNIVERSITY
LEARNING SERVICES

Step 2: Business impacts

This is portrayed by the 'sun' diagram, which allows Carillion to show the linkage between policy, government's objectives and the key sustainability impacts (waste, design, community, etc.). This diagram can be used when considering the impacts of a specific project or contract and has become the Carillion definition of sustainability.

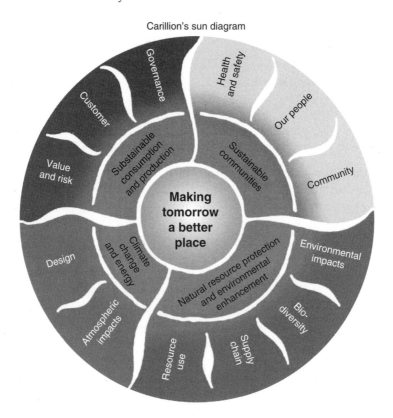

Carillion's sun diagram

Step 3: Strategy model

In 2001, a strategy model was developed to link business objectives to the sustainability objectives. The model was developed with senior team involvement, including the chief executive, as a tool to understand how key performance indicators (KPIs) could be used to improve sustainability performance whilst demonstrating how they deliver business benefit and contribute to the achievement of business objectives.

In 2004, the strategy model and objectives were reviewed to ensure they were based on tangible outcomes to be delivered by 2010. Critically, this demonstrated that it was as important to deliver the intangible issues as it is to deliver hard-nosed cash-backed business objectives. In fact it was seen that community and environmental objectives both enabled, and legitimised, the success of Carillion's business objectives in a process of continual improvement.

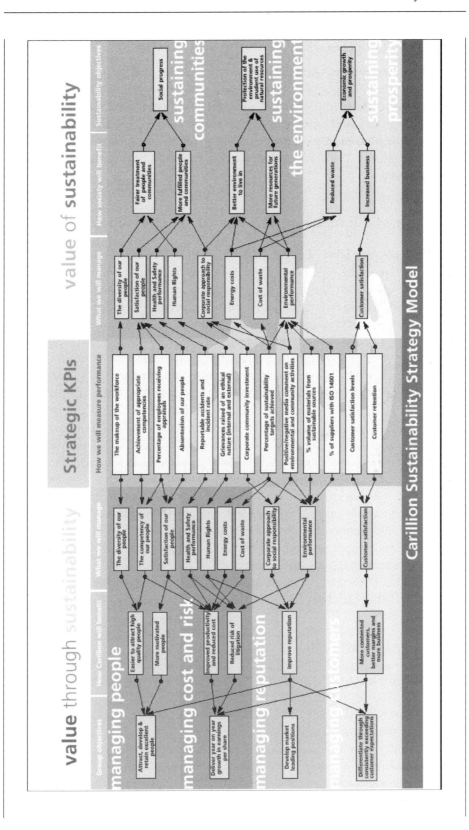

Carillion Sustainability Strategy Model

Step 4: Action plans, communication and training

A suite of action plans provides the necessary focus and management tool to ensure delivery. Throughout these steps Carillion have realised that this can only be successfully implemented with consistent communication of the sustainability policy and aims. It is important that everyone understands their roles and responsibilities to enable the company to become a socially responsible organisation. This is a priority from induction and throughout career development.

Delivery of the sustainability strategy to operations

Carillion has top level accountability to deliver the sustainability strategy, shared between the executive director (environment and health and safety) and chief executive (community, marketplace and human rights), with each business group also having board members responsible for these key issues.

The sustainability committee drives the strategic focus for sustainability, meeting four times a year and drawing on the experience of two respected external advisors. Operationally, sustainability is delivered by the sustainability operations group (SOG) who initiate, capture and communicate good practice. This helps to ensure that the business groups receive the support they required to deliver sustainable solutions. The interaction within and between these groups allows clear communication from strategy to practice.

Each business group delivers the broader sustainability strategy through an integrated management system (IMS), ensuring all potential impacts from businesses operations and projects on the environment and society are effectively addressed. All projects have a project specific management plan as part of the IMS, which ensures projects identify sustainability issues and opportunities in a sustainability action plan (SAP). Once these have been identified it becomes part of the auditing process which enables Carillion and its clients to determine the success in managing sustainability.

Throughout this process, meaningful consultation and dialogue with stakeholders – clients, partners, communities, staff or contractors – takes place.

Development of KPIs

Over the past few years, Carillion has developed its ability to offer sustainable solutions that not only incorporate best environmental and social practice, but also reduce whole-life costs and improve value for money. This has been done by using a model, developed as part of the strategy model (Step 3), which linked sustainability targets to business strategy and financial objectives. The process was supported by analysis of the costs and benefits of sustainable solutions.

By using indicators to monitor and measure its progress, the organisation has benefited from greater innovation, enhanced individual and corporate responsibility and improved product delivery in both construction and maintenance activities. The strategy model demonstrates that by integrating sustainability with business strategy through the use of KPIs, benefits can be gained by businesses and communities.

Part of Carillion's development has been the realisation of the need to, and value to be gained from, comparing its performance against not only competitors in the same industry but also against companies in other industries. Carillion's performance is gauged through its annual sustainability targets consisting of:

❑ Five corporate sustainability objectives;
❑ Supported by nine sustainability targets;
❑ Which in turn were supported by KPIs; and
❑ These aligned with the business strategy using the strategy model, above.

Carillion identified its sustainability impacts in line with the government's key themes for sustainability, which provided a consistent approach and helped the company better understand the concept of sustainable development. This, however, had to be supported by the business case for sustainable development through the quantification of its costs and benefits. The use of sustainable indicators is a key part of this process.

Business can demonstrate to stakeholders that sustainable development is embedded in the control of their activities. However, this can only be achieved if sustainable development is part of the overall business strategy. The majority of business sectors in the United Kingdom are beginning to investigate the benefits of adopting the principles of sustainable development, with the rate of adoption varying within and across business sectors.

In developing the strategy model, and in order to identify clear benefits to be gained from measuring sustainability performance, a review of each KPI was undertaken to identify which business issues could be managed better by using KPIs. A number of indicators were identified as strategic to the business, whereas others helped to deliver business improvements. A third set of indicators, which the company will continue to report as a matter of good practice, were identified as important in demonstrating good corporate responsibility.

Using this approach to review the value of each KPI has helped to demonstrate the link between business benefit, and has shown how KPIs contribute to the delivery of company objectives.

Achievement of business benefit

When any new 'initiative' is introduced to a company, there is inevitably some scepticism from employees, who may well have seen numerous ideas come and go over a period of years. It was, therefore, important to show that the sustainable approach was not a passing fad, but a business imperative, which provided demonstrable benefits for the business, environment and society.

A business case was assembled that comprised project-based case records from which a number of clear benefits were identified. The case records demonstrated that being more sustainable enabled the company to:

(1) Identify social and environmental impacts;
(2) Reduce cost, use fewer raw materials and create less waste, resulting in savings;
(3) Reduce risk and minimise the risk of prosecution;
(4) Improve relationships with customers;
(5) Improve relationships with the community;
(6) Create more effective supply chain management; and
(7) Achieve greater employee motivation.

Benefits were therefore realised for all stakeholders and the whole supply chain, including client, constructor, supplier and maintenance contractor through to the end-user and local communities.

The company has now made sustainability a major part of its developing business strategy. It has mapped sustainability onto the strategy model, and has also used the model to help demonstrate the business case for its supply chain performance measurement process and its waste management strategy.

A reputation for addressing sustainability issues has resulted in increased competitiveness, enabling the company to bid for, and win, a number of multi-million pound contracts where sustainability credentials were one of the deciding factors in awarding the contract. The life-cycle costing model used at tender stage has helped to identify significant savings that could be achieved over 25 years by investing a modest upfront cost with paybacks in less than two years.

Next steps

The sustainability strategy will continue to be driven by the Corporate 2010 Strategic Objectives and Targets and the Carillion Standards. Each business group has developed sustainability targets, cascading the corporate targets throughout the business and setting targets for their own business needs:

2010 objective	Target
Social progress	
To be a recognised leader for the positive impact we have on the communities with which we work	1. Produce guidance for, and roll-out community engagement strategy, with particular focus on schools and local authority roads
	2. Demonstrate 15% improvement in occupational health and safety against Carillion excellence model

2010 objective	Target
Protection of the environment and prudent use of natural resources	
To demonstrate peer group leadership in delivering solutions to the environment and prudent use of natural resources.	3. Produce guidance for, and roll-out biodiversity action plans
	4. Each business group to identify cost of waste and establish action plan to reduce (the waste)
	5. Define strategy for:
	(a) marked reduction in car option CO_2 emissions, i.e. steer the car list to lower CO_2 vehicles
	(b) achieving smarter vehicle mileage, i.e. without negative business impact
Economic growth and prosperity	
Our sustainability approach significantly contributes to us reducing costs by 10%	6. Roll-out sustainability cost framework on new health PFI projects, principally through life cycle and supply chain measures
	7. Complete roll-out of sustainability toolkit to all relevant staff in business development and work winning in each business group
Sustainability is a critical factor in 20% of our orders for new work	8. Improve satisfaction of our people:
	(a) Deliver internal communications plan
	(b) Measure satisfaction of our people using quarterly surveys and great debate to demonstrate improvements
	(c) Revitalise Carillion values
Our approach is a significant contributing factor to at least 50% of our new starters	9. Produce an integrated strategy to reduce 'churn' for recently hired monthly employees

3.2   Master planning and real estate

The inclusion of sustainability considerations into the front-end stages of facilities and major projects can provide significant benefits in cost savings, innovative design and saleable/rentable premises. A step change is required to provide the necessary measures including ring-fenced capital expenditure, objectives for delivery into the project and into operation and the development of sustainability within the brief.

The role of FM at this stage should be significant. FM will become the body who will ensure the measures identified are bedded into the running of the facility, but more importantly are required in the operation of the facility.

Section learning guide

This section describes the incorporation of sustainability within facilities at the feasibility and briefing stage. Decisions made at this stage can influence the level of environmental and social measures incorporated:

❑ Define the initial needs;
❑ Development of sustainability objectives at the briefing stage;
❑ Decision-making points and where sustainability initiatives can be made;
❑ Assessment of leased properties against sustainability criteria and mechanisms to check applicability to objectives; and
❑ Provision of a green lease.

Key messages include:

❑ Decisions should be made at the beginning and carried forward into the rest of the project; and
❑ Sustainability objectives should be set and incorporated alongside those of cost and timescales.

3.2.1   Introduction

The inclusion of sustainability at the briefing stages of a project is performed in part, but rarely classified as such. Decisions on whether to remain in existing premises through refurbishment, or increasing the number of workstations or move to a new facility are commonly considered coupled with the provision of car parking spaces for non-central facilities. However, aspects related to flood risk are not often covered. The benefits which can be achieved have been mentioned previously and include those summarised below:

❑ Reducing facility operating costs through improved energy and water efficiency within the decision-making process;
❑ Setting capital cost provisions for sustainability measures based upon agreed and fixed payback mechanisms;
❑ Choosing property locations which reduce liability related to climate change impacts affecting asset value – storm damage, flooding, legal fines and clean-up;

❑ Incorporating the need to future proof and managing change to deliver facilities that are flexible in the in-country marketplace;
❑ Improving the internal environmental conditions and workplace to increase productivity, optimised internal conditions and retention of staff; and
❑ Involving the community to support the local economy, services and people.

Typically, there are four main triggers for a change in facility, whether new or existing facilities are involved:

❑ Acquisitions and mergers – Before a company is acquired, a review of the portfolio should be undertaken from a sustainability risk perspective as part of the due diligence process incorporating the financial processes. Primarily, the review will identify liabilities and contaminations which exist within the portfolios and measure these as a cost. The review of environmental liabilities is commonly performed as a follow-up once the acquisition has taken place rather than beforehand, often leading to additional unnecessary costs;
❑ Expansion in number of staff – Considerations at this stage include whether a new building is required, or if people can be managed in the existing buildings through reorganisation and different work densities;
❑ New locations/regional hub sites – Site locality is pre-defined in general but the optimum location can be chosen based upon incorporating sustainability principals in the procurement process. If a new building is required to be built or if one could be rented. If a leased property, considerations at this stage include whether the sustainability criteria associated with the property should also be examined;
❑ Consolidation of properties – This could involve either a reduction in properties, or consolidation to a new site. Considerations at this stage should be based upon location, transport and reorganisation of the workplaces and these should all incorporate sustainability criteria.

3.2.2 Briefing / option appraisal

The first activity is the development of the sustainability priorities and commitments in the context of the facilities needs and requirements and the corporate responsibility policy. The value FM can provide at this early stage, in an advisory role, includes intimate knowledge of the organisation, ways of working, goals and culture.

The development of these priorities has been covered in Section 2.1.5. The priorities will be focused upon the corporate sustainability strategy, how the strategy integrates with the project and business needs and the identification of the overall budget resources and life cycle capital funding.

Typical questions to ask and resolve at the briefing/option appraisal stage are provided below to support the identification of the sustainability requirements for the facility. These questions can be used to test the various options available and provide some of the knowledge to deliver an informed decision on the chosen option. Other questions may apply depending upon the specifics of the project.

(1) Are sustainability objectives developed through a CR policy?
(2) Is there a budget to implement sustainable designs, either as a specific budget or incorporated into the project budget as a whole?
(3) What is the planned occupancy life cycle of the facility to calculate payback timeframes?
(4) Have planned life-cycle costs been set?
(5) Have renewable energy consumption targets been set?
(6) Is existing infrastructure in place – gas, electricity, drainage?
(7) Is the facility near a transport hub and are there any limitations on car parking?
(8) Will any development be on, or adjacent to, a brownfield/contaminated site?
(9) Will the facility be on an area at risk of flooding?
(10) Are protected ecological habitats or species within 1000 m?
(11) Identify existing community provisions and outstanding or future needs of the community in consultation with community groups and representatives.
(12) What will be the local regeneration impacts (local jobs and supplies)?
(13) Are there likely to be any protected buildings or public opposition to any changes to the facility?

The matrix below has been developed to support the identification of the sustainability requirements for the facility from which the various options can be assessed (Table 3.2). The 14 areas of sustainability have been described against typical objectives an organisation may set. The matrix should be used at a senior level meeting to openly discuss the areas of focus for the project.

For each of the various headings, the benefits and disadvantages should be captured to enable an informed decision on the main objectives to be taken forward. The environmental and economic sections are self-explanatory reflecting their specific areas, governance refers to a management decision affecting the building.

Following the identification of the various objectives within the project, these should be developed into the project scope and brief describing how these areas will benefit or affect the project, and the objectives required to be delivered. These objectives should be outcome based requirements and therefore easily demonstrable and measurable.

At the end of the briefing stage the following activities should be covered:

❑ Confirm the sustainability objectives for the project;
❑ Confirm the specification for scope and requirements of the project; and
❑ Establish the project programme and budget allowing time and money for sustainability measures to be identified and included.

Table 3.2 Sustainability briefing matrix

Sustainability issue	Environment	Governance	Economic
Management Emissions to air	BREEAM/LEED CO$_2$ impacts Other emissions	Management system (ISO 14001) Mitigate project impacts	Budget availability – ring-fenced Threshold emission limits
Land contamination Workforce occupants	Contaminated land Awareness	Liability management Thermal, lighting and acoustic controls	Remediation cost Satisfaction and performance
Local environment and community	Nuisance, e.g. noise, traffic	Community involvement Planning restrictions	Use of local suppliers/contractors
Life cycle of building/products	Climate change, e.g. flood	Building life/occupancy	Refurbishment Life-cycle cost
Energy management	Energy use Equipment selection	Energy efficiency targets Operating maintenance	Tax rebates Renewable energy
Water management	Water use reduction Pollution control/leaks Local watercourses	Water efficiency	Tax rebates Cost avoidance
Use of resources	Cradle-to-cradle resources	Recycling targets	Recycling revenue
Waste management	Minimisation and recycling	Procurement and segregation	Waste avoidance
Marketplace	External benchmarking	Supplier relationship Stakeholder engagement	Supplier relationship Compliance
Human rights			
Biodiversity	Use of indigenous plants	Biodiversity action plan	Ecological features
Transport	Green transport plan Planning permission	Local transport restraints	Incentives to travel Additional transport

3.2.3   Planning and sustainability impacts

Aligned with the identification of the objectives is the capture of sustainability risks as part of the briefing stage. Risks will arise from sustainability legislation, planning requirements and associated project risks. The aim is to identify potential issues based upon the objectives and build these into the concept design to reduce any impacts. Typical examples will include the provision of renewable energy such as solar and wind, or the provision of car parking spaces. Again, FM has a role through the provision of knowledge and advice.

Risk assessment processes are designed to ensure development needs are met without creating unnecessary risks to people and properties. In view of climate change and the growth in facilities, it is all the more important that planning decisions take into account:

❑ protection and enhancement of the natural and historic environment, the quality and character of the countryside, and existing communities;
❑ ensuring high quality development through good and inclusive design, and the efficient use of resources;
❑ ensuring that development supports existing communities and contributes to the creation of safe, sustainable, liveable and mixed communities with good access to jobs;
❑ mitigating the future effects of climate change including flood risk, increased solar intensity, variable weather patterns and warmer summers; and
❑ the protection of the wider countryside and the impact of development on landscape quality including the conservation and enhancement of wildlife species and habitats and the promotion of biodiversity.

Environmental impact assessment

A recognised mechanism to evaluate these risks is environmental impact assessment (EIA). EIA is a systematic process of evaluating the environmental effects of a proposed development, during construction, operation and after use, leading to the preparation of an environmental statement (ES) to accompany an application for planning permission (Figure 3.3).

There are a variety of reasons to perform an EIA including:

❑ From a legal standpoint, EIA may be a necessary component of a planning application;
❑ A surge of recent legal challenges has made it increasingly important for developers and property owners to carefully consider the need for an EIA at an early stage in the planning process. The implications of failure to correctly follow the process can be severe if a planning permission is subsequently challenged because an ES has not been produced. It can result in costly legal battles and protracted project delays including quashing of permission potentially; EIA can serve a positive role in the development of a master plan for urban regeneration and overall design of a proposal;

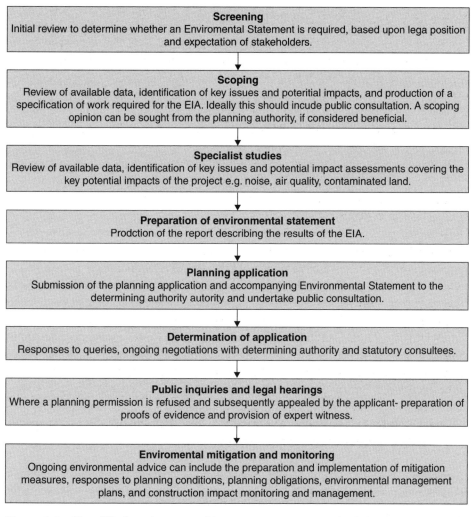

Screening
Initial review to determine whether an Enviromental Statement is required, based upon lega position and expectation of stakeholders.

Scoping
Review of available data, identification of key issues and poteritial impacts, and production of a specification of work required for the EIA. Ideally this should incude public consultation. A scoping opinion can be sought from the planning authority, if considered beneficial.

Specialist studies
Review of available data, identification of key issues and potential impact assessments covering the key potential impacts of the project e.g. noise, air quality, contaminated land.

Preparation of environmental statement
Prodction of the report describing the results of the EIA.

Planning application
Submission of the planning application and accompanying Environmental Statement to the determining authority autority and undertake public consultation.

Determination of application
Responses to queries, ongoing negotiations with determining authority and statutory consultees.

Public inquiries and legal hearings
Where a planning permission is refused and subsequently appealed by the applicant- preparation of proofs of evidence and provision of expert witness.

Enviromental mitigation and monitoring
Ongoing environmental advice can include the preparation and implementation of mitigation measures, responses to planning conditions, planning obligations, environmental management plans, and construction impact monitoring and management.

Figure 3.3 Simplified environmental impact assessment methodology.

❏ Helps to promote the scheme to the authorities and other interested parties; and
❏ Pre-empts or counteracts objections.

EIA may be mandatory or can be performed voluntarily to proactively identify environmental and social risks as part of the overall project. The EIA framework provides the methodology to evaluate these risks based upon developing the baseline situation and then identifying the potential significant effects of the proposed development. Measures to avoid, offset or reduce adverse effects (mitigation measures) can then be implemented. These measures may result in the design being altered or potentially processes may be identified to control and manage effects once the development is operational.

Whether this is a major project, where the EIA is mandatory, or a smaller extension or refurbishment, where a voluntary EIA is undertaken, this process of determining the impacts of the proposed project is very useful. The findings from this exercise provide detailed information on key sustainability issues which need to be managed and can be incorporated into the project scope and briefing.

The main ES document typically runs to 100–200 pages with illustrations, including all the issues identified at the scoping stage, such as noise and vibration, air quality, transport, landscape and visual. Technical appendices and a non-technical summary also comprise part of the ES documentation. The ES demonstrates to interested parties, and importantly the decision makers, that all significant relevant environmental effects have been identified and appropriate mitigation measures proposed to reduce adverse effects.

3.2.4 Pre-occupation assessment

An increasing number of organisations are managing all or part of their property portfolios through leased office and technical centres dependent upon immediate requirements. The days of large property portfolios for companies to manage have gone, epitomised by a decision made by many blue chips in the 1990s to focus upon core business, which did not include the purchase and management of land or property.[11]

Fast-moving sectors such as telecom, financial and IT are very much investment based and are increasingly at the hands of the market. The short-term leasing of property allows a flexible and direct business driven approach to managing growth, smoothing out the peaks and troughs and enabling the business to focus upon core activities.

This is placing an increasing emphasis on business needs for leased property. When a suitable property is ultimately chosen there are a number of questions and factors which must be considered which will influence this decision. However, the needs of the workforce as far as the workplace are generally overlooked. The process usually revolves around the facilities manager needing to move into a given property to ensure it meets the requirements of the business within a fixed timescale. Significant changes to the workplace environment to meet legal, occupancy and cultural needs must be made at a high cost prior to the move-in date.

A structured model for organisations to identify their sustainability workplace objectives prior to the occupation is provided (Table 3.3). This will provide not only a closer match with requirements, a standardised workplace environment across the portfolio, but also identify additional costs as required and manage these effectively as part of the whole project budget. When assessing properties, the use and location of real estate can impact on sustainability. These must be identified and prioritised in order to make informed decisions about the type and location of property based upon the identified sustainability objectives.

Table 3.3 provides a series of questions with tick box responses to elicit a response from landlords and property owners. The range of responses run from

---

[11] *Changes in Property Portfolios*, International Facility Management Association – http://www.ifma.org.uk

Table 3.3   Property pre-occupation sustainability assessment

Management

Is information about specific energy and water usage, waste and recycling volumes and associated costs available?	No data collected ☐	Partial data available ☐	Whole building data available ☐	Tenanted area data available ☐
Does the building have any recognised environmental management system or certification (e.g. ISO 14001, BREEAM, LEED, etc.)?	No ☐	Internal system ☐	BREEAM / LEED ☐	Certified EMS ☐
Have awareness/education programmes been run?	No ☐	No, but allowed ☐	Yes, limited ☐	Yes ☐

Emissions to air

Is there a regular refrigerant maintenance plan in place?	No plan ☐	No plan, alternative control ☐	Basic system ☐	Full system ☐
Are low emissions furniture, carpets and finishes used?	No ☐	Planned ☐	Yes, limited ☐	Yes ☐

Land contamination

Is the site located on or next to contaminated land?	No study performed ☐	Yes ☐	Yes, limited contamination ☐	No ☐

Workforce occupants

Have workstations been checked to ensure they are suitable for the people using them?	No checks ☐	Planned ☐	Yes, some ☐	Yes, all ☐
Is the workplace adequately ventilated with fresh clean air from windows or mechanical ventilation?	No, poor ventilation ☐	No, improvement planned ☐	Yes, adequate ☐	Yes, good ☐
Is there adequate surface illumination from electrical lighting sufficient to enable working practices?	No, insufficient lighting ☐	No, partially met ☐	Yes, adequate ☐	Yes, task lighting ☐
Do staff have sufficient natural light?	No, none ☐	Within 10 m ☐	Within 5 m ☐	Less than 5 m ☐

Local environment and community

Are local suppliers and contractors specified when tendering services?	No ☐	Limited contracts ☐	More than half of contracts ☐	Yes, all services and materials ☐
Have there been any complaints received from the local community?	Varied complaints ☐	One type of complaint ☐	One complaint received ☐	No complaints received ☐

Life cycle of property/products

Are services and equipment procured on a best value basis?	No ☐	Limited contracts ☐	More than half of contracts ☐	Yes, all services and materials ☐
Is full-life costing included in the best value calculations?	No ☐	Limited contracts ☐	More than half of contracts ☐	Yes, all services and materials ☐

Energy management

Does the building have a building and energy management system (BEMS) installed?	No system ☐	No system, alternative control ☐	Basic system ☐	Full system ☐
Does the building have any special energy efficient building features, e.g. night cooling, passive stacks, chilled beams, geothermal?	No features ☐	Yes, some features ☐	Partial natural ventilation ☐	Fully naturally ventilated ☐
What percentage of energy savings have you made in the past 5 years?	< 5% ☐	5–10% ☐	10–20% ☐	>20% ☐

Continued

Table 3.3 *Continued*

Water management

Are water consumption meters installed?	None installed	No, could be installed	Some installed	Installed per tenant
	☐	☐	☐	☐
Does the building have a water conservation programme, e.g. low-flow plumbing fixtures, aerators on tap nozzles, collection of rainwater, use of grey-water?	No programme	Planned programme	Some programme implemented	Yes, full programme
	☐	☐	☐	☐

Use of resources

Are cleaning products selected which have a reduced environmental impact?	No	In development	Yes, some selected	Yes, all selected
	☐	☐	☐	☐
Is recycled paper (at least 80% post-consumer waste and chlorine free) in general use?	No	No, planned	Yes, some uses	Yes
	☐	☐	☐	☐
Is there an up-to-date and maintained register of all hazardous substances used on site?	No	No, planned	Yes, some materials	Yes
	☐	☐	☐	☐

Waste management

Is there an existing recycling programme or space available for one to be established?	No	No, but allowed	Yes, limited	Yes
	☐	☐	☐	☐
Are separate bins available for recyclable items?	No	No, but allowed	Yes, limited	Yes
	☐	☐	☐	☐
Are opportunities for office equipment recycling available – electronic equipment, furniture?	No	No, but allowed	Yes, limited	Yes
	☐	☐	☐	☐
What percentage of office waste is recycled?	< 20%	20–50%	50–70%	>70%
	☐	☐	☐	☐

Marketplace

Are cost effective services provided – to meet with service level agreements?	No	Limited contracts	More than half of contracts	Yes, all services and materials
	☐	☐	☐	☐

Human rights

Are the needs of disabled people met within the facility?	No, not met	No, planned	Partially met	Yes, fully met
	☐	☐	☐	☐

Biodiversity

Is there a biodiversity action plan in place inside or external to the facility?	No plan	Ad hoc approach	Basic action plan	Full plan in place
	☐	☐	☐	☐

Transport

How many on-site car parking spaces are available per occupant?	>1.0 per person	0.7 to 1.0 per person	0.4 to 0.7 per person	<0.4 per person
	☐	☐	☐	☐
How long does it take to walk to the nearest public transport location, e.g. bus stop, underground, train station?	>20 minutes	10 to 20 minutes	5 to 10 minutes	<5 minutes
	☐	☐	☐	☐
Does the site have a green transport plan?	No plan	No plan, alternative control	Basic system	Full system
	☐	☐	☐	☐
Does the site have video-conferencing facilities?	No plan	No plan, alternative control	Basic system	Full system
	☐	☐	☐	☐

'no information available' or 'no action performed' through to 'best practice'. Following completion of the assessment, properties can be ranked against each other through a scoring mechanism, either as a uniform system, or by applying weightings based upon the objectives defined at the briefing stage. Using this method, the most sustainable property can be identified to make an informed decision.

3.2.5   Green lease

Most leases do not promote environmental or social improvements stating fixed energy and waste costs as part of the service charge.

Sustainability issues should be incorporated within lease agreement through negotiation of a 'green lease' where appropriate. Whilst this may be provided as a wish list, which not all premises may be able to provide, informed decisions can be made based upon the provisions which can be made. The power of such questions will also help to drive the market forward, as more tenants request such lease information, the market will be encouraged to provide the measures described below.

A green lease may include the following:

❏ Sharing the cost of recycling schemes with other tenants;
❏ Sharing the cost of energy efficient equipment with other tenants;
❏ Incentive for energy saving and recycling initiatives in the calculation of service charge rates;
❏ Clause protection for any environmentally friendly equipment or building alterations within dilapidation clauses;
❏ Are noise restrictions placed to limit when works can take place?
❏ Does vegetation/trees have to be maintained as part of the lease?
❏ What is the payment provision for utilities – gas, electricity and water?
❏ Is the service charge paid as a percentage, or flat rate?
❏ A request to define number of disabled parking spaces;
❏ Statement that preference is for direct meter payment. Where not available, what rebates are available to incentivise energy efficiency returns?
❏ Statement requesting how waste disposal costs will be managed and availability of reports;
❏ Statement requiring information on the system specification, controls and zoning, how they influence energy consumption and the ability to modify heating and cooling patterns within the building;
❏ Statement requesting services in the building or nearby providing for staff including food service, luncheon clubs, health clubs, hotels and other retail facilities.

3.3   Design

The design of a facility, whether new, an extension or major refurbishment, has arguably the greatest sustainability impact of any decision made in the life cycle of the facility. At this stage, a step change in environmental and social performance can be incorporated into the new facility.

A number of factors need to be incorporated within the design process for the Facilities Manager – primarily ensuring a clear specification for the occupants and the physical building itself is developed to provide a framework around which the design can be developed. Unfortunately, this approach is not as common as it sounds, with the role of FM traditionally to comment on designs already developed. Many facilities are designed at the initial stages without taking into account waste minimisation during the construction process, and energy provisions such as renewable generation.

Section learning guide

This section provides the incorporation of sustainability criteria into the concept and detailed design of facilities covering the requirements of the facility and the occupants:

❑ Setting of occupancy requirements within the design criteria;
❑ Development of design targets and incorporation within the design;
❑ Sustainability performance tools such as LEED and BREEAM;
❑ Designing in waste avoidance; and
❑ Procurement of building materials with a recycled content.

Key messages include:

❑ Operational specifications for the facility should be set and agreed during the concept design phase;
❑ Maintenance of the facility will be determined at the design stage;
❑ Up to 3% of facilities' costs can be saved through designing in waste minimisation.

3.3.1   Introduction

It still costs more to construct green buildings, but the financial benefits of green-building design are more than ten times the average cost premium. A study in October 2003 by the California Sustainable Building Task Force found that constructing a certified green building costs on average about 2% more than a traditional building of the same size. But the extra cost yields a tenfold saving over 20 years through lower energy and water bills, reduced waste disposal costs and increased productivity and health of workers. For example, an extra investment of $100 000 to install green features into a $5 million project would save at least $1 million during the next 20 years, the report stated. Five years ago, green buildings were unusual, expensive and it was unclear what the benefits were with a cost premium anywhere from 5 – 15% more. Now, materials and design processes have become standardised, and more people know how to design green buildings. An integrated design and commissioning process is the most cost-effective. If you take a conventional building and add piecemeal green technologies or design strategies, you end up with a substantially more expensive building. The more expensive green buildings are those that had late change orders. The report concludes that constructing green buildings to the LEED gold level (equivalent to BREEAM Very Good) makes the most financial sense.

The best time to consider how to minimise required energy is before a building is constructed. Architects and engineers collaborating on a building's orientation, lighting system and materials promote the use of sunlight as more of an asset and less of an energy drain. In some cases, designing a building in the shape of a rectangle instead of a square and orienting the longer sides in certain compass directions can lessen the surface area exposed to the hottest sun of the day. Lowering the amount of energy used by lighting does not mean employees work in the dark. Fixtures combine direct lighting with indirect lighting that deflects light off of the ceiling onto a work area. Such systems cut the number of required watts per square foot and can reduce demand by 50%. That's significant when considering that lighting often makes up 40% of a building's energy load. Equally important, the lighting is comfortable and utilitarian to the building's users. Using windows and skylights as the primary source of illumination, and supplementing the natural light with a dimming system or low wattage bulbs can also reduce power consumption. A building's envelope itself can save energy. Window glazing with light-reflective coatings reduces summertime heat, and exterior shading devices allow good light to enter and keep intense heat out.

Investments in high performance buildings pay for themselves. Often it's not a question of spending more money, it's spending money in the right place. On a 355 000-square-foot building in Minneapolis, US, installing high performance windows, though more costly, can reduce the need for heating and air conditioning, reducing demand, and providing smaller specification units. Many new buildings don't perform up to expectations, even though they meet relevant energy codes with energy costs varying as much as 35 – 40% for supposedly similar neighbouring buildings. A large percentage of these inefficiencies are designed into the buildings, based solely upon up-front costs rather than life-cycle costs.

The Cooperative Insurance Society (CIS) service tower, a Grade II listed building, already dominates the Manchester skyline and is the tallest office building in the United Kingdom outside London. New cladding will now make it the largest vertical solar cladding project in Europe and something of a tourist attraction. The building is more than 40 years old and the small mosaic tiles that clad the service tower need replacing. These solar panels are the ideal solution, protecting the tower from the elements, enhancing its appearance and generating significant amounts of renewable energy. Three sides of the 400 foot tower will be clad creating 180 000 units of renewable electricity each year. Although this won't be enough to run the entire building – with over 4000 employees inside – it will significantly reduce the amount of power it needs to draw from the national grid.

Green building has also become a marketing tool. Commercial real estate agents are starting to meet tenants' demands for an environmentally friendly building with natural ventilation – a place where the windows can actually open. After all, a worker in a building with no noxious chemicals is a happy, more productive worker. Of course, only a handful of well-publicised buildings are really green. Most clients are wary of the fact that environmentally conscious constructions can cost more up front, even if savings are promised in building operations over the long haul.

Post-occupancy evaluations of buildings and their surroundings will soon measure physical activity and personal interactions by occupants. In other words, instead of providing sterile lobbies with elevator banks leading to cubicles, buildings will be expected to maximise opportunities to use the stairs, to walk around, and to gather in social spaces – see Section 4.3.

3.3.2   The role of facilities management in design

The design of a building has a major impact and influence on the operational performance of a building – that much people agree on. When this does not happen – whether because designers do not understand the remit, or the operators cannot manage the facility – it is the occupants and the finances that suffer. There are a number of reasons for this gap, which emphasise the role of FM and the end-user at this stage:[12]

(1) Inadequate communication through the delivery of the facility from:

   (a) Client requirements not successfully being implemented through the designs to end-user stage;
   (b) Buildings are not handed over with sufficient process information back-up and training to enable on-going operation to the design intent;
   (c) Facilities managers are involved too late in the design process to provide much influence.

(2) Different agendas.

   (a) Client brief does not always capture the end-user requirements;
   (b) Architects design buildings without enough end-user input;
   (c) End-users have high requirements that are unrealistic and un-maintainable;
   (c) Facilities managers focus on practical and maintainable systems for the end-user.

(3) Lack of understanding.

   (a) Decisions are made without consideration of their long-term maintenance impacts;

---

[12] Cousins, J., Butler, T. and Shah, S. *What designers need from facilities managers*, British Institute of Facilities Management Conference, March 2005.

(b) Facilities managers are left to maintain systems which have not been efficiently implemented or commissioned;

(c) Facilities managers are not integrated early or thoroughly enough in the design process to understand how the building was designed to operate.

(4) Timescale constraints.

(a) Process and operational systems get compromised, over-specified and may not be linked effectively;

(b) Design decisions are rushed and not thought through to understand consequences;

(c) Buildings are handed over too quickly without thorough systems checks.

The role of FM currently is predominantly through the project manager and one step removed from the client and architect (Figure 3.4), limiting the ability to influence design parameters and set end-user requirements. Decisions are often made without necessary feedback and risk problems during operation. The preferred model represents the facility manager function operating with the client, architect and project manager, with a direct link to the end-user to understand needs and a continual feedback loop. This model enables greater incorporation of end-user operational perspective at the initial stages of design and requires a clear brief to be defined capturing the necessities to avoid later mistakes.

Throughout the design process and particularly at value engineering stage, FMs will need to ensure that initiatives developed to meet the brief are not discarded straightaway. The value engineering process is an important one designed to reduce the cost of construction. The role here is to utilise standardised off-the-shelf products to save time and materials, rather than the traditional removal of technologies.

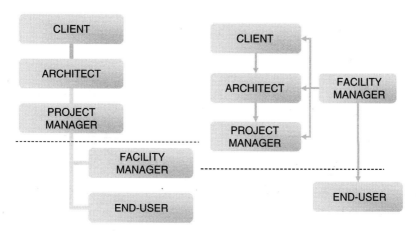

Figure 3.4 Current role of FM in design (left) and proposed role (right) with involvement in the design.

3.3.3   Design targets

The use of targets should not be undervalued in the design process. They help to focus attention on client requirements and should act as a means to deliver a better building. The basis for the targets should be taken from the 'objectives' defined in Section 3.2 based upon the priorities, risks and benefits. From these objectives a series of outcome based targets can be developed at the concept stage to provide the scope for the design team to achieve. The outcome based targets will enable any process or technology to be used allowing greater freedom to the designer, within the confines of the project brief.

Table 3.4 describes the 14 sustainability categories used throughout this book, coupled with units for measurement. Typical indicators are described capturing the main areas where, for example, energy will be used in the facility. These indicators have been developed both for their ease of measurement, as well as their ability to indicate good practice.

3.3.4   Sustainability framework tools

There are two recognised global certification schemes in use. BREEAM (Building Research Establishment (BRE) Environmental Assessment Method) was developed in the United Kingdom in 1990 and is used widely across Australia, Canada and New Zealand. LEED (Leadership in Energy and Environmental Design) was developed in the latter part of the 1990s though has gained credence across the United States, Canada and recently in China and India.

Both schemes are excellent methodologies to incorporate environmental criteria into the design process. However, they are best used as a vehicle to encourage greater sustainability practices at the initial briefing stages, rather than as a direct measure of what to include. Traditionally assessments are performed at detailed design stage to assess number of points and include elements to gain the higher rating if achievable easily. Innovation and benefits are limited which may include many practices, not noted in the methodologies that would provide performance and cost benefits to the building performance.

Both schemes offer a range of benefits, from environmental to financial, including:

❑ Demonstrate compliance with environmental requirements from occupiers, planners, development agencies and developers;
❑ Define 'green building' by establishing a common standard of measurement and provide a means for environmental improvement;
❑ Raise consumer awareness of green building benefits;
❑ Create a better place for people to work and live (occupant benefits);
❑ Achieve higher rental incomes and increased building efficiency;
❑ Promote integrated, whole-building design practices; and
❑ Recognise environmental leadership in the building industry and transform the building market.

Table 3.4 Sustainability outcome performance criteria

Category	Indicator	Factors to include
Management	Certification rating	❑ Achievement to LEED/BREEAM
Emissions to air	CO_2 equivalent/m^2 (NIA)	❑ Major service extracts are designed to minimise external pollution ❑ Intakes for major service routes are designed to minimise internal air pollution, noise and vibration
Land contamination	% brownfield development	❑ Use of brownfield land
Workforce occupants	% occupant satisfaction	❑ Day lighting strategy ❑ Major servicing routes to minimise noise, vibration and pollution
Local environment and community	% sustainable spend	❑ Complies with statutory authority requirements ❑ External/security lighting strategy to minimise night-time light pollution where applicable ❑ Access (disabled, visually impaired, hard of hearing, etc.) ❑ Integration of requirements for key religious and ethnic groups
Life cycle or building/products	Life cycle payback (months)	❑ Efficient construction techniques, prefabrication, etc. ❑ Materials specifications for major elements especially with respect to maintainability and whole-life costing ❑ Cleaning and general maintenance ❑ Building form and orientation to minimise external wind effects
Energy	kWh/m^2 (NIA)	❑ Integrated building services control strategies ❑ Optimise passive design, day lighting, building form and orientation and efficiency ❑ Renewable technologies

Continued

Table 3.4   *Continued*

Category	Indicator	Factors to include
Water	m^3/person	❏ Surface water storage/attenuation measures where required ❏ Protect/enhance existing water features ❏ Collection and use of rainwater/grey water ❏ Set targets for water use minimisation during the construction process and in the finished building
Use of resources	% sustainable spend	❏ Materials specifications for major elements for recycle or re-use
Waste	kg/person	❏ Site waste strategy during construction ❏ Strategy for operational recycling ❏ Set targets on waste reduction during construction
Marketplace	Benchmark score	❏ Performance against peer group ❏ Supplier relationship strategy
Human rights	Benchmark score	❏ No direct impact
Biodiversity	% managed biodiversity space	❏ On-site enhancement and protection of ecological features ❏ Impact on biodiversity, and area of habitat created/retained ❏ Set targets on site ecological enhancement and protection ❏ Site rectification/remediation measures where required
Transport	Number/ workstations	❏ Transport policy to include key issues such as efficient use of cars/delivery vehicles, provision of cycling facilities, etc. ❏ Commercial vehicle movements to set targets in this area during construction

The two schemes are similar in many respects but one notable difference between them is that LEED doesn't have third-party certification. This means that the LEED focus is more on process whereas BREEAM focuses more on performance.

BREEAM – Building research establishment environmental assessment method[13]

BREEAM is the United Kingdom's most widely recognised measure to assess the environmental performance in environmental design and management. The method covers a range of environmental issues, presenting the results in a 'rating' that is understood by those involved in property procurement and management. BREEAM is still very much an environmental tool predominantly focusing upon energy management and carbon dioxide reduction.

BREEAM assesses the performance of buildings in the following areas:

❏ *management:* overall management policy, commissioning site management and procedural issues;
❏ *energy use:* operational energy and carbon dioxide (CO_2) issues;
❏ *health and well-being:* indoor and external issues affecting health and well-being;
❏ *pollution:* air and water pollution issues;
❏ *transport:* transport-related CO_2 and location-related factors;
❏ *land use:* greenfield and brownfield sites;
❏ *ecology:* ecological value conservation and enhancement of the site;
❏ *materials:* environmental implication of building materials, including life-cycle impacts; and
❏ *water:* consumption and water efficiency.

Developers and designers are encouraged to consider these issues at the earliest opportunity to maximise their chances of achieving a high BREEAM rating. Credits are awarded in each area according to performance (Figure 3.5). A set of environmental weightings then enables the credits to be added together to produce a single overall score. The building is then rated on a scale of *pass (25 to 40), good (40 to 55), very good (55 to 70)* or *excellent (70 to 100)*, and a certificate awarded that can be used for promotional purposes.

BREEAM can be used to specify the environmental performance of buildings in a way that is quick and visible in the marketplace, and by property owners to promote the environmental credentials of their buildings.

LEED: Leadership in energy and environmental design[14]

The LEED green building rating system is a voluntary, consensus-based national standard much like the BREEAM. Members of the US Green Building Council

---

[13] Building Research Establishment Environmental Assessment Methodology – http://www.breeam.org/

[14] Leadership in Energy and Environmental Design – http://www.usgbc.org/

Category	Number of credits	Value credit	Maximum score
Management	9	1.67	15
Health and wellbeing	15	1.00	15
Energy	17	0.83	14
Transport	13	0.83	11
Water consumption	6	0.83	5
Materials	11	0.91	10
Land use	2	1.50	3
Ecology	8	1.50	12
Pollution	11	1.36	15
TOTAL			**100**

Figure 3.5   How the BREEAM system works.

representing all segments of the building industry developed LEED and continue to contribute to its evolution. LEED standards are currently available or under development for:

❑ New commercial construction and major renovation projects (LEED-NC);
❑ Existing building operations (LEED-EB);
❑ Commercial interiors projects (LEED-CI);
❑ Core and shell projects (LEED-CS);
❑ Homes (LEED-H); and
❑ Neighbourhood Development (LEED-ND).

LEED provides a complete framework for assessing building performance and meeting green building goals. Based on well-founded scientific standards, LEED emphasises strategies for sustainable site development, water savings, energy efficiency, materials selection and indoor environmental quality. LEED recognises achievements and promotes expertise in green building through a comprehensive system offering project certification, professional accreditation, training and practical resources.

LEED assesses the performance of buildings in the following areas (Figure 3.6):

❑ Sustainable sites – location of site (erosion, flooding, brownfield), green transport and infrastructure;
❑ Water efficiency – water conservation measures, recycling and use of rainwater and grey water technologies;
❑ Energy and atmosphere – commissioning, renewable technologies, and process issues;

Category	Possible points and % of total	
Sustainable sites	14	20%
Water efficiency	5	7%
Energy/Atmosphere	17	25%
Materials/Resources	13	19%
Indoor environmental quality (IEQ)	15	22%
Innovation	4	6%
Accredited professional	1	1%
TOTAL	**69**	**100**

Figure 3.6   How the LEED system works.

❏ Materials and resources – recycling and recycled content materials, provision of recycling space, sourcing of materials;
❏ Indoor environmental quality – ventilation, use of materials which reduce volatile organic compound (VOC) emissions and occupancy comfort; and
❏ Innovation and design process – additional points for exception performance beyond the requirements of LEED.

The rating system is based upon four levels: *certified (26 to 32 points), silver (33 to 38), gold (39 to 51) and platinum (52 to 69).*

Other schemes

There are a number of other national schemes in operation around the world, notably:

❏ Canada – The Assessment Framework and Green Building Tool (GBTool) was originally based on the BREEAM framework;
❏ United Kingdom – design quality indicators provide a methodology to benchmark and assess performance;
❏ Australia – The National Australian Building Environmental Rating System (NABERS) was developed in close consultation with the UK BREEAM team;
❏ Hong Kong scheme – The Hong Kong Building Environmental Assessment Method (HK-BEAM) is also based on BREEAM but has developed since its original inception to reflect local requirements;
❏ EU scheme – HQE is a French-led scheme based more on quality than the other schemes and is unlikely to replace BREEEAM;

❑ Netherlands – Eco-Quantum currently available only in Dutch; and
❑ Switzerland – Eco-Invent.

3.3.5 Designing in sustainability[15]

There is potential for making significant savings by managing waste more effectively. Case studies from live projects have shown average savings of 3% of build costs, or 20% of material on site, and these can be achieved *without* significant investment costs.[16]

When thinking of construction waste, most people automatically think of waste generated during the construction phase of a project. Whilst this is an obvious area to be addressed, it is by no means the only one: for example, it has been estimated that 33% of wasted materials is down to architects failing to design-out waste. Given that 80% of a building's impact is determined at design stage, there is a real opportunity for designers to affect the waste a building creates.

'Solid waste' is not the only form of waste that designers can affect. Other aspects include:

❑ Energy use;
❑ Water use;
❑ Unnecessary consumption of land;
❑ Time;
❑ Lower than planned economic return; and
❑ Unrealised potential from built assets.

There are a number of different aspects of designing a building that can dramatically affect the amount of waste it creates. These include:

Designing in functionality

The aesthetics of a building have a major impact on the wasteful nature of the facility. For example, the shape of a building can be inherently wasteful – in terms of useable floor area, the need to heat unnecessarily large volumes, or non-standard fittings. There is a need to strike a balance between aesthetics and the level of waste within the facility – for example, the heating requirement for a dramatic high-ceilinged reception area could be reduced by ensuring that the building fabric is extremely well-insulated to reduce the demand throughout.

Orientation should be considered from the point of view of both opportunities and challenges. Opportunities include potential reductions in heating costs and impacts by designing-in potential for passive solar gain. Challenges include the

---

[15] Adapted from Sustainability Forum Materials Sub-Group; Sector Sustainability Strategies – Identifying Cross-Cutting Issues, unpublished, 2005.

[16] Constructing Excellence in the Built Environment – http://www.constructingexcellence.org.uk

need to consider the direction of prevailing winds, which can cause damage to a building's structure, thus creating an unnecessary maintenance requirement.

Robustness and durability of building materials is important from the point of view of both financial and environmental costs – 'whole-life costs' include not only the initial cost of a particular material or product, but also the costs associated with maintenance and repair throughout its life, and this is a useful assessment method increasingly required by clients. Unnecessary environmental impacts can be created by the need to replace building components because they have not been appropriately chosen – inappropriate species of timber for cladding.

Simplicity of design may mean that environmental impact is lessened – fewer materials mean that there are more opportunities to re-use or recycle material at the end of a building's life. It may also be more versatile and adaptable to a variety of uses, rather than needing to be demolished and rebuilt, if its form is simple.

Designing in material specification

Specification is an obvious area where savings can be made, both environmentally and financially, from choosing the most sustainable or renewable materials wherever possible. There are other considerations, including choosing the best types of materials for the job in terms of their thermal efficiency and their properties – see Section 3.3.6.

Designing in waste minimisation

The design of a building will impact on the quantities of solid waste produced at the construction stage, and in refurbishment. Efficient use of materials by designing to standard sizes with minimum cutting of materials can be achieved within almost all project types and budgets, reducing waste and disposal costs. They will also keep refurbishments and replacement items cost-effective. Pre-fabricated or pre-assembled materials or modules, particularly large-scale projects, can bring substantial benefits.

The increasing availability of modular buildings across Europe has provided faster, less wasteful, cheaper and safer construction and operation facilities. External opening and finishes are specified alongside internal fittings to enable a complete module to be delivered to site reducing waste by two-thirds. In Germany, many homes are sold in 'flat-pack' at builders merchants involving greater use of timber-framing and plastics. As well as using ultimately more sustainable materials, there are additional benefits through reduced construction costs, lower embedded energy and the facilitating of quicker mass-building techniques.[17]

The main products for replacement or substitution are those containing concrete, steel and certain ceramics by timber and plastics that can be readily used as cheaper and equally effective materials in many structures, where durability and longer term use is factored in.

---

[17] Prefabrication technologies – http://www.fabprefab.com

Designing in efficient building utilisation

Designing buildings that are fit for purpose, flexible and adaptable in layout (for example, open-plan space) and services to avoid major changes in use and enable the facility to change usage from office to retail or hotel for example. Buildings can be designed to be readily converted and refurbished to enable their transfer from one use to another. The design of many old buildings has permitted conversions such as eighteenth century warehouse premises in London and New York.

An increased use of technology in modern buildings to manage the ventilation, lighting and window systems can cause conflicts in the maintenance of the facility, adding significant costs and difficulties in the ability to convert usage.

Designing in character and setting

Making sensitive materials choices and designing in context of the surroundings to reduce impact on the environment are the primary areas of concern. This will include the use of local materials and regional construction methodologies to ensure the design is in keeping with its surroundings. Building to meet and enhance local characteristics will provide a unique facility, and greater asset value. Critically, many of these issues are increasingly required for planning permission and will therefore affect timescales and costs.

Designing in environmental practices

Biodiversity is another area that should be considered at all stages of a project. Designers can make a major contribution to maintaining, protecting and enhancing biodiversity through the design, through the choice of planting and habitats for a range of species. The management of a building also has major impacts on the environment, and designers should take this into account when developing design criteria to which building occupiers will refer.

Designing in business planning

Involving the whole team at appropriate stages of the design can result in better buildability by drawing on varied experience and areas of expertise; this in turn results in less waste. 'Lean thinking' can be applied to design process and activities as well as to the construction phase with impressive results, reducing waste in many areas of activity.

Designing in technology

The development and inclusion of technology to provide more intelligent buildings, integrating a variety of systems can deliver greater level of flexibility and awareness of the buildings operating parameters. Linkage between operating systems such as lighting controls and the building energy management system can be related to access controls within floors to automatically switch off lighting and change the heating and cooling parameters. Much of this level and control will require cabling running above the ceiling or under the flooring.

3.3.6   Procurement – building materials

Many mainstream products commonly used in construction such as blocks, bricks and boards contain significant amounts of material that have been recovered from the waste stream at no extra cost or risk whilst meeting quality standards. For example, different brands of plasterboard may contain between 15% and 99% recycled content. It is not always obvious that new building products contain such material, but, by selecting appropriate products, contractors can be more efficient in their use of material resources without compromising on cost, quality or tight deadlines. Specifying recycled is a strategy to pursue alongside established options for reclaiming and re-using products (such as partitions and ceiling grids). It is a simple way of delivering quantifiable environmental benefits, without compromising on the aesthetic qualities or technical performance requirements that are demanded by your scheme. Targets of 20% recycled content should be set.[18]

Sustainable material process

(1) Collate a list of the main materials to be used in construction, e.g. concrete, glass, aggregate, cladding;
(2) Set targets for sustainable, locally sourced and recycled content materials;
(3) Confirmed standard sizes of, e.g. windows and partitioning, are to be specified;
(4) Calculate locally sourced and recycled content alternatives to traditional sources for the top ten materials (based upon cost);
(5) Specify requirements within tender process, including the need to verify sourcing of materials, e.g. FSC timber; and
(6) Review responses against targets set and confirm chain of custody for sustainable materials such as timber.

Scope should be examined for substitution between production and consumption of materials inputs/outputs from different materials groups, for example, glass, metal and plastics are all used in secondary aggregates and may be as 'fit for purpose' as primary products in many building activities. The substitution of these for primary materials can do much to reduce carbon emissions. Unless re-usable or recycled products comply with recognised product schemes, designers will not use them. It has been an enduring problem in recent years for industries to get their products adequately specified and so be able to market them effectively – the process tends to be lengthy and fraught with difficulties. This is in no small part due to the very large number of new products becoming available for assessment, and also due to the very limited resources available for product testing.

---

[18] Waste & Resources Action Programme – http://www.wrap.org.uk

Building Research Establishment's, (BRE) Green Guide, The American Institute of Architect's Environmental Resource Guide and Green Building Specification provide a remarkably thorough compilation of the environmental aspects of scores of materials. LEED and BREEAM provide some synthesis, incorporating materials-specific standards such as Forest Stewardship Council certification for sustainably-harvested wood, low VOC emissions for interiors, and defining criteria for recycled content or regional sourcing.

A typical list of materials to avoid includes:

❏ Timber from non-forest stewardship certification (FSC) scheme;
❏ Ozone-depleting substance as a coolant;
❏ Insulation materials with a global warming potential above 5;
❏ High volatile organic compound (VOC) content paints, varnishes and finishes;
❏ Lead based paints;
❏ Flooring which will release VOCs;
❏ Asbestos;
❏ Peat and weathered limestone; and
❏ High embodied energy materials.

The primary method for evaluating a material's relationship to the environment is called life cycle analysis (LCA). The analysis covers a product's life cycle from extraction or harvesting through processing and manufacture to use and ultimate disposal or recycling. LCA practitioners must decide where to draw boundaries for analysis, what impacts are considered, and how they are described or quantified. To make it more complicated, a material that may be environmentally benign in one situation might in fact be detrimental in another, if improperly used. For example, it might be possible to specify recycled content insulation that has fewer VOCs and has a lower embodied energy than a conventional product, but if it has a lower R-value than a competing petrochemical product it may result in greater environmental impact over time. Operating energy use usually exceeds embodied energy many times over.

Each stakeholder in building design, construction, operation and occupancy has a stake in how our industry addresses these vital issues. Manufacturers should be responsible for providing the public with accurate data and full disclosure. The best practices of information design should help package this information in the most accessible way. This goal should be balanced with economic factors, and that creation and implementation should be as easy and inexpensive as possible. Give the market full information, and let the market choose what to use in the built environment.

3.3.7   Renewable and low carbon technologies

The vast majority of energy consumption in a facility is during its operation, so any reduction in operational energy consumption through efficient design will

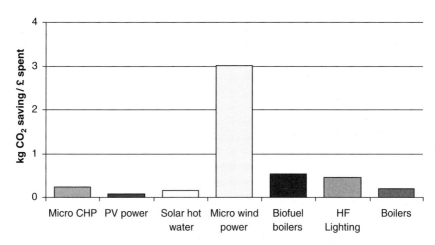

Figure 3.7 Annual carbon dioxide savings per implementation cost (£).

significantly reduce costs. Passive technologies such as increased thermal mass, orientation of the building and insulation can drastically reduce the level of heating and cooling required for any facility.

To supplement these passive measures, renewable technologies can be incorporated into the design of the facility. There is a level of resistance to take up technologies such as micro wind turbines or photovoltaics due to cost pressures, particularly capital cost barriers – the long payback and high up-front investment required for the more visual solutions like photovoltaics is a significant barrier.

Incorporation of renewable technologies does not have to cost significantly more than traditional construction. Incorporation into the concept designs to maximise technologies does not add greatly to the capital costs. Targets for renewable energy should be set for any facility over $1000\,m^2$, given that the size will justify any increase in capital cost with short-term paybacks. In many cases, grants or tax cuts are available to offset capital costs either wholly or for a period of time. The return of carbon dioxide against the investment varies greatly for the various technologies (Figure 3.7), with micro wind turbines by far the most viable. This translates into the cost effectiveness of the various technologies commonly available.

Rainwater harvesting is common for toilets, and new technologies provide the re-use of water in toilets from recycling treated sewage effluent. Coupled with water efficient appliances and life styles, demand can be significantly reduced.

Solar PV costs in Japan fell by 75% between 1994 and 2004, during which time there was a 35-fold increase in installed capacity, over which time central government subsidies fell from 50% to 10% over the same timeframe and will be phased out completely this year. In the United Kingdom six panels or 1 kWp of solar PV on an average home built to latest building regulations will reduce carbon emissions by more than 20%. Guaranteed by the manufacturers for 20 or 25 years, they will last far longer. In the case of building integrated solar PV, these installations also offset the costs of other traditional roofing and facade materials. This is of particular

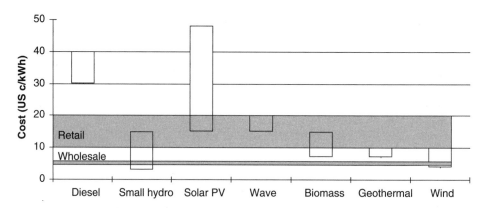

Figure 3.8 Renewable energy technology costs against traditional retail energy prices. (Source: adapted from Shell Renewables report 2004 and Building Options for Renewable Energy, Carbon Trust 2003.)

relevance to the commercial sector. Building designs will therefore need to adapt so that they can readily integrate such systems in efficient and cost-effective ways, e.g. PV panels as a cladding material.

The chief drawback of renewable technologies is their cost compared to conventional energy sources. With global retail prices at between 10 and 20 US cents per kWh, few schemes can compete on price. The cost of generating electricity from wind is low compared to that of solar or wave. The cost of electricity from conventional sources is currently low, however changes in security of supply and costs are enabling renewable sources to compete directly (Figure 3.8). Shell Renewables' report from 2004 highlighted that wind turbines and biomass boilers, of suitable size and location, could close the gap on conventional sources if energy prices continue to rise.

Table 3.5 describes a variety of low-carbon and renewable technologies available on the market, the mechanisms of how they operate and how to maximise their output. In the final column, a series of issues which may preclude the use of these technologies is provided. For the installation of any of these technologies, a feasibility study and subsequent study should be performed to understand the practicalities of installation, risks in place and additional costs, with the aim of producing a cost with a plus or minus 10% variable.

Case Study: Heelis Central Office, Swindon – The National Trust

Drivers for the central office project

The National Trust's central office directorates were spread across six different and dispersed sites in the south of England with some departments having three or four locations. They needed to bring the staff together at a single location within easy reach from the existing locations.

We set out to be innovative, not pioneering. Now working under one roof (One Trust) we can:

❏ work smarter and offer a modern, integrated service to our members;
❏ realise significant economies in running costs;
❏ enable faster communication within the National Trust, providing improved team working;
❏ reduce environmental impact from travel and buildings; and
❏ lead by example and practice what we preach.

Sustainability is a key issue in the National Trust's philosophy. The Heelis building had to meet the high quality and high level benchmarks for sustainable design. We also wanted an open-plan environment to ensure good links and cross communication between departments. The use of day lighting and natural ventilation demonstrates how a building can be architecturally enticing and an outstanding place to work whilst being highly sustainable with low running costs.

So what makes Heelis different?

By starting with a 'blank sheet of paper' in a new building, the Trust has designed the most effective way for us to work better, drawing on the valuable expertise of staff and consultants to learn the best practice from each of them. All aspects of office service information, from how the mail system works to the on-line booking of meeting rooms is accessible and works extremely efficiently.

Heelis is an effective 'fit for purpose' office building. It is also a safe and healthy building. Good architecture, good internal layout and attractive landscaping enhance the working environment. The internal environment is also improved by the use of natural daylight wherever possible and natural ventilation. We have taken care to develop consistent ways of working, supported by high standards of furniture and equipment. It is a place staff want to come to – it is visually interesting, warm and friendly with a good sense of community and team spirit. There is a great social space, excellent food for staff, well-equipped tea points and business centres, and there are lots of places to visit at lunchtime – this gives staff the opportunity to stretch their legs and rest their minds away from the working environment. The design encompasses open-plan and non-hierarchical space, which aids internal communications, and there are peaceful, varied and excellent meeting spaces.

A natural environment

For most of the year the building is naturally ventilated. This is achieved through a combination of automatically controlled windows and a series of roof vents. These vents open into raised snouts on the roof, which act as umbrellas to keep out the rain and as chimneys drawing air into the building from the perimeter and courtyard windows. Some snouts have low-velocity fans to encourage air movement on still days. During the winter, when opening all the windows causes significant loss of heat and energy, a simple mechanical ventilation system feeds fresh air through the raised floor maintaining the internal air quality. Another sustainable feature is to use the warm extract air to heat the incoming fresh air again reducing our operational costs. The system relies on the thermal mass by using the concrete ceilings to absorb heat from the office space during the working day and maintaining a comfortable working environment. During the hot months the external vents are opened during the night to purge heat from the mass and cool the structure for the following day. Much like historic churches and many buildings in hot climates the exposed thermal mass absorbs heat and provides free cooling to balance the diurnal temperature changes.

The north-facing roof lights across the entire building reduces to a minimum the need for electric lighting. When it is needed an automatic control system adjusts the level of artificial light delivered in response to external conditions. Movement sensors ensure that lights are switched off in unoccupied areas.

Natural resources

Timber from our woodlands and wool from Herdwick sheep grazed on Trust farmlands help make Heelis a unique environment. In the atrium, at the heart of the building, the wall cladding and stairs are made entirely from wood from National Trust estates. The carpet was developed for the building using undyed Herdwick yarn mixed with a nominal amount of nylon and carbon fibre to create a commercial quality but natural carpet within the costs and with all the benefits of the standard man-made product.

There are over 1200 photovoltaic panels (solar energy) mounted on the roof, the major part of the cost of which were met from a government grant. The panels collect solar energy, which is used to generate electricity either for immediate use in the building or to be sold back to the grid. A display in reception shows the amount of energy we generate and the carbon emissions this cuts down. We expect the installation to provide about a third of the electricity we use in the building, producing a major and increasingly important operational saving as energy costs continue to escalate. The panels also act as shading to the north-facing windows, an ingenious solution to the problem of solar glare within the building on mid-summer days.

An excellent BREEAM rating

Heelis is designed to minimise environmental impact not only in use but also in its construction and eventual demolition. All design decisions were aimed at using materials sustainable in their production, use and disposal as well as to minimise energy consumption. This philosophy went so far as to take action to avoid traffic movements wherever possible. Heelis is expected to produce only 15 kg of CO_2 per m^2 annually, compared to 169 kg for a typical air-conditioned office and 57 kg for a typical naturally ventilated office. The scheme received an excellent rating – the highest possible – from the BREEAM (Building Research Establishment Environmental Assessment Method) scheme.

Facilities management overview

Recognition of the FM element within the project at an early stage has contributed to achieving not only a benchmark building for sustainable construction but also one that is sustainable in operation. Indeed the building was fully operational from day one following a single weekend move from five existing offices. The in-house facilities lead was appointed to the project steering group in the summer of 2003 at the commencement of Cat B (fit-out design), when the building was literally an empty shell. The operational facilities management team was appointed in early 2005 for a building opening date of 4 July 2005.

The FM remit from day one was not only to input into the fit-out design but also the wider operational issues that collectively informed new ways of working in conjunction with the IT and HR functions within the Trust. Recognising that sustainability went further than bricks and mortar and had to embrace staff retention, work life balance and green travel.

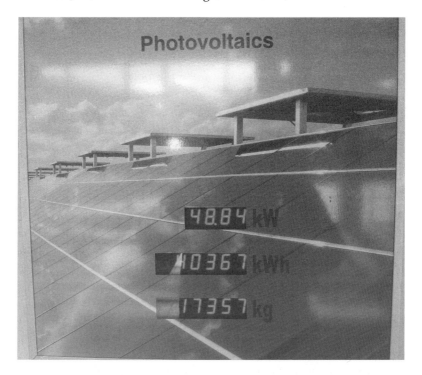

Taking the role of the informed client in design team briefings, early involvement meant the future operation of the building had a strong influence on the architectural and engineering solution. Generic solutions and space standards were challenged specifically in back of house areas resulting in some previously unconsidered areas such as a dry recycling store and drying room for staff clothing. A process that as design progressed went into the detailed design and layout of operational areas and questioned how maintenance access would be achieved. This role became more informative with time as the Trust's culture, structure and ambitions were assimilated into the project, informed by open consultation and debate with a group of super users representing each of the existing locations and functions.

The Heelis project was a precursor to 'insourcing' the core facilities team. The unique arrangement of public (NT shop and café) and staff accommodation informed by the NT property model resulted in an operational facilities manager (repairs and maintenance, cleaning, security, post room) and a visitor services manager (reception, meeting rooms, switch board) supported by a team of facilities co-ordinators and customer care assistants respectively. Service partners (cleaning and associated services, repairs and maintenance, manned security guarding) were identified, following a comprehensive tender process, aimed at identifying not only best value but also innovation and sustainable awareness. This resulted in contractors being available for the mobilisation of Heelis from practical completion.

Communications

It was important that we not only showed all staff what the building would look like but also how it would work. In addition to the wider concerns with relocating and job security, there was a lot of negativity and uncertainty especially with the open-plan arrangement amongst staff at all locations, many of whom had always worked in a small cellular environment. We had to persuade our staff that Heelis would be unlike any other office they'd visited as well as the benefits of open-plan working. This process commenced 18 months before the move with a series of all staff presentations and visits to the site, culminating in detailed presentations three months before the move. In these final few months we launched a comprehensive on-line induction that informed staff how Heelis would work and also discharged our requirement to communicate key health and safety information, completed with an automated email acceptance on receipt of which we issued staff access cards. The final piece was a detailed welcome pack, the Heelis handbook that expanded on the induction providing detailed instruction on how to access systems and facilities, e.g. how to print, how to book a room, how to book travel tickets. In addition to the technical we also had to communicate expected behaviours or etiquettes appropriate for open-plan working, which dealt with some basic considerations, e.g. using voicemail when away from your desk, keeping ring tone low, not walking the office whilst making a mobile phone call. These were supported by a series of charters, which clearly identified facilities and staff roles and responsibilities in ensuring the best possible service.

Procurement

Fixtures and fittings were procured with the support of consultants, Buro 4 Business Services. The choice of workstation was a critical element of the project; strict criteria were set for product design, construction and use in operation balanced against sustainable construction and transport. The final space plan

and specification was informed by the building design to optimise performance, e.g. screens were kept low to avoid shadowing from roof lights.

Operational achievements

It was important that the in-house facilities team were both accessible and
accountable to all staff. We also wanted to reduce the dependence on administration and empower all staff to take control of their daily work life. Our
solution was to introduce online meeting room and rail travel booking systems
that improve efficiency and reduce costs. The requirement for the project to
demonstrate overall operating savings of £500k+ p.a. means we need to ensure
maximum benefit for minimum cost.

LIVERPOOL JOHN MOORES UNIVERSITY
LEARNING SERVICES

Unwanted print was a major source of wastage within the existing central offices; due to their cellular nature, there were a high number of personal printers. This arrangement resulted in a high dependency on printed matter. Through a co-ordinated programme of education and technology we have achieved paper savings of 40% or 20 000 pages per week:

❑ Nine multi-functional devices (print, copy, fax) set up in six business centres;
❑ PIN code system so nothing prints until chosen – no accidental prints;
❑ User sees menu of their jobs so can still delete unwanted items;
❑ Default to duplex printing;
❑ Jobs sent to print deleted from the queue after 24 hours;
❑ Flat screen monitors encourage easier reading from screen, i.e. no need to print;
❑ General awareness of waste reduction goals; and
❑ Minimum storage on site for paperwork.

Wastage also led us to re-evaluate the need for under desk waste paper bins, but in doing so we wanted staff to consider their actions. The solution following an exhaustive search identified a European range of recycling pods with separate sections for glass, plastic, paper and general waste. We also recycle cardboard with an onsite compactor, toner cartridges, mobile phones and lamps in accordance with best practice.

3.4   Construction and commissioning

The construction of the facility, either as a new-build or major refurbishment, will deliver the agreed designs. The role of FM during this phase is limited, other than to provide an on-going monitoring role to ensure health and safety issues and environmental and community impacts are effectively managed.

Table 3.5 Renewable and low carbon technology comparison matrix

Technology	Energy conversion and utilisation	Key requirements for optimising viability	Key aspects affecting implementation
Photovoltaics	☐ Daylight to electrical energy ☐ Enables electrical power for utilisation within the building	☐ Unshaded access to daylight for PV panel location ☐ Southern orientation for PV panel ☐ 30°–45° elevation of tilt for PV panel	☐ Suitability of roof structures to mount PV panel arrays ☐ Urban location influences potential planning restrictions
Photothermics	☐ Solar radiation to thermal (heat) energy ☐ Enables hot water heating for ablutions and catering use	☐ Unshaded access to the sun for solar panel location ☐ Southern orientation for solar panel ☐ 30°–45° elevation of tilt for PV panel	☐ Suitability of roof structures to mount solar panel arrays ☐ Urban location influences potential planning restrictions ☐ Central provision for hot water production required (not individual localised heaters)
Micro wind turbine	☐ Wind to electrical energy ☐ Enables electrical power for utilisation within the building	☐ Unsheltered location for micro wind power location	☐ Suitable external areas to mount micro wind turbine ☐ Urban location influences potential planning restrictions
Biofuel heating	☐ Organic vegetable matter (tree wood) combustion converted to thermal (heat) energy ☐ Enables heat generation for utilisation in building	☐ Access to sufficient, reliable local resource of renewable fuel, i.e. wood chip or pellets	☐ Suitable location for biofuel boiler location ☐ Suitable location for fuel storage ☐ Sufficient provision for fuel delivery ☐ Urban location influences potential planning restrictions (boiler flue discharge position)

Continued

Table 3.5 *Continued*

Technology	Energy conversion and utilisation	Key requirements for optimising viability	Key aspects affecting implementation
Geothermal	❑ Natural ground source heat sink conversion to higher grade heat or cooler via electric heat pump ❑ Enables increased efficiency in operation of mechanical heating and cooling systems (heat pump)	❑ Access rights for extraction of local groundwater ❑ Geotechnical survey to establish local suitability of site	❑ Existing sites may not be in suitable locations to access most optimum groundwater aquifers ❑ Space for boreholes or closed loop exchangers may be limited ❑ If no mechanical cooling, utilisation may be limited
Micro CHP	❑ Utilisation of waste heat from simultaneous electrical power generation at point of use ❑ Enables localised production of heat and power – with savings in CO_2 emissions from use of waste heat	❑ Year-round requirement for heat utilisation	❑ Suitable location for CHP unit adjacent to existing boiler plant ❑ Domestic hot water services to be served from space heating boilers if possible to maximise year round heat demand
Rainwater harvesting	❑ Utilisation of rainwater from surface run-off from roofs for non-drinking purposes, i.e. flushing WCs ❑ Enables less mains water utilisation, saving in CO_2 emissions from mains water treatment process	❑ Significant utilisation of non-potable water required ❑ Suitable location for storage tank required	❑ Suitable location for storage tank may be difficult to achieve without incurring major cost associated with burying tanks and re-instating existing ground cover finishes ❑ Alterations to below ground surface water drainage may be necessary

Often, the time period leading up to the handover of a facility is under pressure with all parties working to meet set deadlines. This can lead to shorter than necessary handover of information and poor knowledge by the end-user to manage the facility. Careful preparations are required through the construction programme to understand how the building has been designed and educate the FM team on its operational requirements prior to commissioning and handover.

The first stage is to establish clear definitions and guidelines on what the sustainability features of the building are and ensure the constructor is fully aware. This may be in the form of a section in the tender and subsequent briefing, or through a more formalised training programme. Many construction staff will move from one project to the next carrying out the same culture and therefore education is critical to ensure they are aware that projects are different.

Section learning guide

This section reviews the environmental and social measures during construction and commissioning. Primarily this involves the information and knowledge transferred from the design and construction teams through to the FM team in demonstrations and training sessions:

❏ Construction environment and social management plans;
❏ Site waste management plans; and
❏ Commissioning and handover.

Key messages include:

❏ Involve FM early in the handover and commissioning process; and
❏ Set strong environmental and social measures during construction.

3.4.1 Construction environmental and social management plan (CESMP)

The environmental and social management during the construction process is delivered through a CESMP, by providing an overview of the key issues and actions. The methodology is similar to the environmental management system ISO 14001 detailed in Chapter 2 (Section 2.4). The requirements of the CESMP include:

❏ Identification of significant environmental aspects and impacts;
❏ Identification of environmental controls and mitigation measures;
❏ Setting environmental objectives and targets;
❏ Demonstrating compliance with relevant legislation and other relevant guidance;
❏ Devising emergency plans to deal with potential environmental incidents; and

❏ Stating roles, responsibilities and communication links for key members of the team.

The CESMP will remain a 'live' document throughout the construction works, documenting environmental and social risks. Commitments or other relevant information will be kept to provide a record of advice to those responsible for subsequent work on the site.

The process starts with establishing objectives and targets, which have been developed for relevant environmental and social aspects of significance during the review process. They are designed to limit any deterioration in the existing environment and, where possible, lead to the improvement of conditions for each of the environmental and social issues, which may be affected by the project. Performance against the targets will be assessed at regular intervals during progress meetings. Key stages are topping out, first fix, second fix, commissioning, practical completion. To support the CESMP, a training and awareness programme should be implemented to meet the particular needs of the project, staff responsibilities and requirements to meet the CESMP.

Typical risks which have the potential to significantly impact the environment include:

❏ Construction noise and effects on local residents;
❏ Ground contamination;
❏ Aggregate dust/excavation dust;
❏ Mud on roads (if temporary unpaved surfaces are present);
❏ Waste handling/segregation;
❏ Fuel spills/oil spills from site chemicals storage or equipment;
❏ Discharge to sewers;
❏ Temporary access for the site works, and maintaining access for the community; and
❏ Implementation of training, including topics on spillage procedures, noise, contaminated land.

3.4.2   Site waste management plan (SWMP)

A site waste management plan (SWMP)[19] provides a framework to improve environmental performance, meet regulatory controls and reduce rising costs of disposing of waste. Adopting a site management approach based around an effective SWMP can bring many benefits including:

❏ Better control of risks relating to the materials and waste on site;
❏ Provides a demonstration of waste management and cost and risk control;
❏ Compliance with legislation; and

---

[19] *Site Waste Management Plans: Guidance for Construction Contractors and Clients*, Voluntary Code of Practice, UK Department of Trade and Industry, July 2004.

❑ A framework to make cost savings through better management of materials supply, materials storage and handling and better managing your waste for recovery or disposal.

There are nine important steps to producing a SWMP:

Step 1: Identify who is *responsible* for producing the SWMP and ensuring that it is followed. Different individuals may be responsible during the planning stages and the site-work stages. Individuals must know what they are responsible for and have sufficient authority to ensure that others comply.

Step 2: Identify the types and *quantities of waste* that will be produced at all stages of the work programme/plan.

Step 3: Identify waste *management options* based on the waste hierarchy, on- and off-site options and managing any hazardous wastes produced.

Step 4: Identify waste *management sites* and contractors for all wastes and set-up contracts emphasising legal compliance.

Step 5: Carry out any necessary *training and on-going awareness* of in-house and sub-contract staff so that everyone understands the requirements of your SWMP.

Step 6: *Plan* for efficient materials and waste handling taking into account any site constraints.

Step 7: *Measure* how much waste and what types of waste are produced and compare these against your SWMP. These figures should be recorded on the datasheet.

Step 8: *Monitor* the implementation of the SWMP and update if circumstances change.

Step 9: *Review* how the SWMP worked at the end of the project and identify learning points for next time.

A series of questions has been developed to support the production of a SWMP before starting on site, during the operational phase and post-completion to identify lessons learnt:

Procurement

❑ Has a careful evaluation of materials been made so that over-ordering and site wastage is reduced?
❑ Has full consideration been given to the use of secondary and recycled materials?
❑ Is unwanted packaging to be returned to the supplier for recycling or re-use?
❑ Can unused materials be returned to purchaser or used on another job?

Operation

❑ Has responsibility for waste management planning and compliance with environmental legislation been assigned to a named individual at both main contractor and identified sub-contractors?

❏ Has an area of the site been designated for waste management, including segregation of waste?

❏ Have targets been set for the different types of waste likely to arise from the project?

❏ Have measures been put in place to deal with expected (and unexpected) hazardous waste?

❏ Has disposal of liquid wastes such as wash-down water and lubricants been considered?

❏ Has agreement been sought from the sewerage company for trade effluent discharge?

❏ Have opportunities been considered for re-use of materials on or off site?

❏ Have opportunities been considered for on- or off-site processing and re-use of materials?

❏ Have you considered what are the most appropriate sites for disposal of residual waste from the project?

❏ Are there opportunities for reducing disposal costs from waste materials which may have a commercial value?

Monitoring and management

❏ Are selected waste materials segregated to allow best value to be obtained from good waste management practices?

❏ Are containers/skips clearly labelled to avoid confusion?

❏ Are any checks made that waste is received at the intended site?

❏ Is implementation of agreed waste management procedures monitored?

❏ Are reports regularly produced regarding waste quantities and treatment/ disposal routes, and on costs incurred?

Reporting

❏ Has a final report of use of recycled and secondary materials, waste reduction, segregation, recovery and disposal, with costs and savings identified, been completed?

❏ Have key waste management issues been considered for action at future projects?

3.4.3 Commissioning and handover

Good installation, commissioning and handover of a facility are critical to its on-going operational performance. A major part of this process is the transfer of knowledge and information between the design and construction team and the team involved in managing the facility on an operational basis. Commonly, this transfer is limited and particularly for speculative developments, information is in the format of paper files. Lack of attention has led to many facilities performing poorly and being managed incorrectly against their original design intent.

At this stage the performance of the building should be thoroughly tested to ensure it meets with the anticipated design. This is also the time to fully involve the FM team who will be running the plant and equipment, cleaning the facility or providing the landscaping. Typical areas to check include:

❑ Air leakage rate/air tightness and pressure tests;
❑ Insulation of the facility and ventilation;
❑ Specific fan power;
❑ Sub-metering and remote monitoring;
❑ Demonstration to users;
❑ Daylight linked lighting;
❑ Night cooling strategies;
❑ Primary plant set-up;
❑ Use of cleaning chemicals; and
❑ Planting and landscaping management.

Successful commissioning will lead to the safe and reliable operation as per manufacturers' recommendations and maximise the operational life of the equipment. An additional requirement is the efficient use of energy through ensuring the facility is correctly balanced, sufficiently air-tight and insulated. A provision should be made to enable the FM team to carry out familiarisation tours and training sessions ready for the handover.

At the building handover, the knowledge on how the facility operates should be transferred from the design team to the FM team:

❑ A building log book for the operator;
❑ Full documentation on the commissioning including a comparison between the design intent and the commissioned building to ensure compliance;
❑ Operating and maintenance manuals;
❑ Overview of the overall design and control strategy and building services operation;
❑ Efficient plant operation during seasonal changes, start up and shut down and out of hours; and
❑ Operation of equipment and controls.

Alongside the operating and maintenance (O&M) manuals, a building log book should be provided to help reduce energy and maintenance costs. The log book should provide information in a ready-to-use and easily read format capturing the critical information on the building. Key information should include:

❑ Schedule of supply meters and sub-meters, highlighting the fuel type, location, identification and description;
❑ Energy rating of the building and zones if applicable and how performance can be calculated from the meter readings;

❑ Overview of the building design, commissioning and handover performance;
❑ Strategic understanding of how the building operates in a non-technical manner.

The log book should act as the single point of reference for the building linking information from the O&M manuals and other critical documents. As such, it should be maintained and be updated as the building changes, becomes fine tuned and more data become available. Alterations to the document should be made, signed and dated, with copies of the original information maintained. The log book also needs to be stored in a safe and accessible location.

3.4.4 Handover checklist

At the building handover, the O&M manuals will contain the vast majority of information concerning the on-going operation of the building. Coupled with the non-technical building log and a users' building manual, sufficient information should be in place to effectively manage the building. Table 3.6 provides a checklist of items during the building handover.

Critical areas are the provision of information within the non-technical and users' manual since these two documents will provide the bulk of information to staff about how they can support the operation of the building.

Liaison with the design team should continue post-handover as the building starts to bed-in, as a number of issues are likely to arise where design input, and in many cases, problem solving is necessary to rectify them. These lessons are important not only for the FM team, but also for the designers to incorporate into any future designs. It is recommended holding a series of informal meetings during this time period, with a formal meeting at the end of the year to close out any issues. In addition, a further visit is recommended after two or three years to again assess performance and ensure the facility is still operating at its optimum.

Satisfaction surveys are also important both for the end-user to take into account any issues affecting performance, but also of the client, the FM team and the designers. Together these will help to inform the gap between the desired and received facility from the client and FM perspective. It will also identify both good and bad points from the design team involved.

3.5 Refurbishment, fit-out and project management

The refurbishment of the space, whether following the handover of the shell and core of a facility, or throughout the life cycle of the facility as it changes has a significant impact upon sustainability. During each stage there are a number of materials purchased to replace those disposed of, involving packaging and transportation directly, alongside labour and materials indirectly. The provision of '100% recyclable' products, particularly furniture, is commonplace, however there are few places where this is not feasible due to the volumes and difficulties of returning the equipment.

Table 3.6 Typical handover checklist for the facilities manager to ensure satisfactory acceptance of the facility. (Source: adapted from Building Research Establishment Post Occupancy Evaluation.)

Performance of the building	
Adequacy of building commissioning records verified	Record sheets for all items in commissioning schedule signed by competent person
Satisfactory methodology in place for those items known to require post-handover commissioning	Contract documents held by project manager include a methodology for resolving items on agreed list. This details activities of key players, and time frames for action.
Completion of post-handover commissioning	Records of any post-handover commissioning undertaken are signed off by the facilities manager
Performance of the building assessed against design intent	Records of the performance assessment against the design intent signed by the facilities manager
Indoor air quality testing performed, e.g. VOC levels	Records of testing and compliance to occupant safety levels signed by the facilities manager
Technical information and training	
Adequacy of BMS and controls commissioning records verified (Month 3)	Record sheets for all items in commissioning schedule signed by competent person and placed in O&M manual
Satisfactory remedial actions in place for those controls and BMS items requiring post-handover commissioning	Contract documents held by the project manager include a methodology for resolving items on agreed snag list, which details actions, responsibilities and time frames for action
Completion of post-handover and seasonal commissioning	Records of any post-handover commissioning are signed off by the facilities manager and kept in the O&M manual
Adequacy of O&M manual verified at handover	The contents comply with the requirements laid down in the contract. To be signed off by the project manager and placed in the O&M manual

Continued

Table 3.6 *Continued*

Schedule of facilities staff training approved (to include technical staff, security staff, and cleaning staff, etc. as appropriate) and formally performed	Confirmation that a schedule of facilities staff training has been developed and performed as per the original contract documentation. This should provide details of the schedule and actual training performed for: the programme; numbers to be trained; means of testing and certification; and any other items required. Detailed requirements confirmed as appropriate by the design and building management teams and signed off by the facilities manager
Suitability of O&M manual reviewed by occupier	(a) Initial comments by the facilities manager on the suitability of the O&M manual at the time of handover have been received (b) Further comments have been received from the facilities manager and facilities staff after training is complete and experience has been gained (c) Revisions made to O&M manuals as appropriate, i.e. error or inadequacy and signed off by the facilities manager
Confirm that appropriate operational and maintenance procedures are in place	(a) Early discussions have taken place between the client and the design team to confirm requirements, which have now been acted upon (b) Discussions are currently ongoing to put in place appropriate procedures
Non-technical information and training Formal arrangements made to train new technical and building management staff	Facilities manager confirms that adequate arrangements are in place and are recorded in personnel files
Formal arrangements made to train new staff occupying the building	Facilities manager confirms that adequate arrangements are in place
Feedback sought on the adequacy of the training given and remedial measures taken	Facilities manager confirms that a system of post-training assessment is in place; that any feedback received is recorded and that remedial action is taken
Schedule of staff training documented and performed	A detailed programme of training has been prepared, which includes any elements agreed within the original contract documents, and has been agreed with the facilities manager. Signed attendance records are held and managed by the facilities manager.

Forward management	
Plant drawings and building documentation safely stored for ease of viewing and updating, together with a secure archive	Facilities manager confirms that arrangements are in place including: Drainage drawings Location of oil/diesel store Meter and sub-meter details Cleaning requirements Transport details
Ease of access to emergency information verified	Facilities manager confirms that emergency information is clearly highlighted, and emergency procedures are properly included in staff training
Formal responsibility for updating of building information allocated	Facilities manager confirms that adequate arrangements are in place. All changes made, and the reasons why, are recorded.
Suitable monitoring and targeting programme established for environmental and social impacts, e.g. occupant satisfaction, transport, water, energy	Facilities manager confirms that this is in place, accompanied by adequate sub-metering, and guidance from the project team on expected first and subsequent year's performance
Appropriateness of FM budget and staff verified	Facilities manager confirms that any pre-existing budget and resource forecast has been considered in the light of building performance
Mechanisms established for appropriate routine observational checks by facilities management staff	Facilities manager confirms that mechanisms have been put in place to carry out routine observational checks. The method to be used is updated as required.
Client – lessons learnt	
Formal satisfaction surveys of the client team to obtain both positive and negative building performance feedback	Facilities manager confirms that a system is in place and that records of discussions held and actions taken are held with the building log book
Formal end-user post-occupancy satisfaction survey undertaken to obtain both positive and negative building performance feedback	Facilities manager confirms that a system is in place and that records of discussions held and actions taken are held with the building log book

Continued

Table 3.6 *Continued*

Formal interviews carried out with facilities staff/maintenance team to obtain both positive and negative building performance feedback	Facilities manager confirms that a system is in place and that records of discussions held and actions taken are held with the building log book
Lessons learnt from the building in relation to the suitability of original brief over the first 12 months' occupancy formally recorded	In the case of a repeat client a formal consideration of the implications for the original building brief has been made and the record taken passed to the appropriate client representative
Widely promoted end-user contact point or help facility available to log and to assist with complaints	Facilities manager confirms that this is available and that records of all complaints logged, with the remedial actions undertaken
Feedback provided to staff on building operation	Provision of a users' building handbook to provide information on the building operation and means to feedback
Formation of building user group to provide continuous feedback	User group has been formed with a clearly defined remit and reporting mechanism

Design – lessons learnt

Client aware of contact details for relevant project team members should problems occur	Contact details for all involved parties are located within the building log book
Special arrangements for problems have been put in place	Explicit processes have been documented on when and how to involve the project team in collaborative problem solving
Design team involvement in client feedback activities	Client satisfaction and design team satisfaction surveys performed
Formal wash-up sessions planned between the client and the project team	(a) A more frequent informal system of review between the facilities manager and the design team; (b) 12 months post-occupation formal meeting between the facilities manager and the project team to consider the issues arising; (c) 24–36 months post-occupation formal meeting between the facilities manager and the project team to consider the issues arising

Through effective procurement of materials and project management organisations, much of this 'wastage' can be removed.

Section learning guide

This section discusses the cradle-to-cradle philosophy and how it can be incorporated into the purchasing of materials and furniture. This involves specifications laid down to project managers and post-project assessments:

❑ Cradle-to-cradle philosophy;
❑ Refurbishment and fit-out impact assessment; and
❑ Sustainable materials and furniture.

Key messages include:

❑ Look to re-use and recycle any unwanted furniture and project materials – companies will take them for a limited fee.

Case Study: An energy-efficient house – Refurbishing Housing

The concept of the energy-efficient house often conjures up images of wooden huts, or a door in a hillock covered by grass surrounded by wind turbines and photovoltaic sources. Energy efficiency doesn't necessarily show up on the outside – it can look just like any other house.

Dave Hampton, a life long advocate of energy efficiency and 'greenness' moved to his current home in Buckinghamshire, UK, in 2001 with designs on making it a low carbon emission home from day one. It was, he says, the toughest project he's ever had to manage. (Apart from getting plumbers and builders to follow exact instructions, he had a consultancy to run and four children to care for!)

The vision is becoming self-reliant on energy and achieving a zero carbon household. Dave's belief is: 'Sustainability is a big field and if you crack the carbon bit of it all the other bits start to fall into place. I see carbon as the key for unlocking the whole jigsaw to sustainability . . . and it's more urgent.'

The most visible feature is probably the solar water system, where the water is 100% solar heated in summer. The five-bedroomed detached house boasts a solar thermal panel that helps save at least £100 on natural gas and a tonne or so of carbon dioxide emissions. In addition there is a 'whole house heat recovery' system that cleverly maintains the property's fresh air levels by extracting moisture while retaining the heat. Other features include a log-burning stove, very high levels of insulation, ultra low flush-volume toilets, efficient lighting and a condensing gas boiler, plus a device that calculates and displays how much electrical energy is being used and at what cost – real time.

Case Study: London Studios – Building Design Partnership (BDP)

Formerly the Cannon Brewery's Fermenting House, BDPs London Studios were originally built in phases between 1895 and 1898. Built on load-bearing brick, with a facing of soft red brick, the building was designed by the architects Bradford and Sons. Inside, special arched concrete floors were used to withstand the weight of the heavy equipment, supported on a network of cast-iron circular columns and wrought iron compound girders.

There were storage and packing rooms in the basement and on the ground floor, yeast rooms on the first floor, skimming rooms on the second floor and laboratories on the third floor with fermenting and tank rooms – all with tiled walls to keep the interior cool and hygienic. The upper floors housed the coolers and refrigerators and the roof held a 36 000 gallon cast-iron water tank. An iron and steel bridge linked the fermenting house with the main brewery building, to its west, now largely redeveloped.

After the brewery's closure in 1955, the fermenting house was converted to a bonded store and offices, and was later used as warehousing and by Christie's for the display of modern art. It was restored and refurbished in 1999–2000. BDPs own designers created three major studio floors, break-out areas, formal meeting rooms and a café and a restaurant, all with flexible framework to maximise use of the space. Numerous original features have been retained including the exposed steel columns, vaulted ceilings and original glazed ceramic brick-walls. BDPs design approach has been largely to retain existing surfaces and introduce any new elements as a contrast to the original characteristics.

The basement is used to concentrate major paper storage and leave floors as open as possible, supported by a document management system. Workspace is based on $6 \times 2\,m^2$ platforms shared by 6–8 people. A wireless network allows additional flexibility including the use of the café and restaurant for workspace. The flexible ground floor is already used as a venue for all types of events.

The building houses 371 staff and boasts a low energy consumption due to the largely naturally ventilated and high thermal mass space providing good control of temperatures in both summer and winter. Natural lighting has been provided through utilising the existing large windows thus reducing the need for artificial lighting, supported by meeting rooms located in the central core.

3.5.1   Introduction

The traditional handover of a building, either as a shell and core requiring full fit-out, or occupation of a tenanted location requires a level of refurbishment either as a category A or B fit-out. Table 3.7 provides a list of the common fit-out requirements from both category A and category B covering both new and refurbished premises. It should be noted that for most of these items, sustainability criteria will play a significant role in the choice and life-cycle costs of the associated items.

3.5.2   Refurbishment and fit-out

One of the greatest activities which has uncontrolled environmental impacts is through project works varying from minor refurbishment and churn, to building extensions. Ultimately sustainability considerations should be reviewed within the design, construction and operational phases of a project to minimise the impact. The reasoning behind reviewing the impacts at each phase is to provide mitigation of the known risks, but to identify new risks apparent from more information provided. In addition, the inherent nature of projects is that they constantly change, and therefore the initial design may not be the one chosen to go ahead, resulting in different risks.

It is important that relevant legislation and corporate, trade or industry standards are incorporated into the process. This provides the basis for ensuring that the environmental impacts can then be managed and minimised throughout the duration of the project. To facilitate this, sustainability should be included as an agenda item within meetings throughout this time period with particular focus on the design concept, build and completion stages to identify and incorporate environmental best practice.

Table 3.7   Category A and B fit-out criteria

Category A fit-out	Category B fit-out
Emulsion paint/coverings to walls	Floor finishes upgrade or modifications
Raised floors and carpets including skirtings	Suspended ceilings upgrade or modification
Suspended ceilings including fire barriers	Internal partitioning to form cellular offices etc. including wall coverings
Lighting and emergency lighting	Doors, frames and ironmongery in cellular offices
Power installation, floor boxes and earthing	Reception desk
Heating installation	Kitchenettes including cupboards, work tops, equipment and service supplies, drainage, etc.
Ventilation installation	General office signage (internal and external)
Air-conditioning installation	Blinds to external windows
Building management control systems	Alterations to lighting and power installations to suit office fit-out layout
Fire alarm system including detectors and sounders	Alterations to heating, ventilation and air-conditioned installations to suit office fit-out layout
Builders' work in connection with services	Alterations to fire alarm system to suit office layout
Statutory signage	Alterations to sprinkler installations to suit office layout
Sprinkler installations	Telecommunications cabling
	Data cabling
	Public announcement system including speakers and cabling
	CCTV and monitoring systems
	Security systems
	Hose reels and fire extinguishers
	Specialist lighting
	Vesda protection systems
	Standby generators and associated plant
	Builders' work in connection with services
	Office furniture, notice boards, etc.

Each project should be reviewed for sustainability measures at the briefing stage. A generic assessment template is provided in Table 3.8 which incorporates the 14 sustainability categories present during the project activity itself, and also any changes upon completion of the project.

The assessment should include consideration at the design stage of using products and materials that will minimise the impact on the environment during the

Table 3.8 Template for assessing sustainability risks during projects

Sustainability category	Relevant (Y/N)	Design issues	Build issues	Operation	Performance measures
Management					Certification rating
Emissions to air					CO_2 equivalent/ m^2(NIA)
Land contamination					% brownfield development
Workforce occupants					% occupant satisfaction
Local environment and community					% sustainable spend
Life cycle or building/ products					Life-cycle payback (months)
Energy management					kWh/m^2(NIA)
Emissions to water					m^3/person
Use of resources					% sustainable spend
Waste management					kg/person
Marketplace					Benchmark score
Human rights					Benchmark score
Biodiversity					% managed biodiversity space
Transport					Number/ workstations

whole of their life cycle. A vast amount of information has been documented about the choice of materials and specifically furniture, and it is constantly changing.

At each stage of the project, the table is updated, providing the high level information, with underlying documentation and areas to minimise the underlying risk. The first process is to identify whether the impact is relevant to the project – a chiller replacement project will involve hazardous materials and energy consumption, but is unlikely to have any great impact on material usage.

It is important that any dust or particulate matter that may be generated as part of a project to refurbish or construct a building is carefully controlled and managed. Any release into the atmosphere will impact upon local residents and the community, and can be acted on through the local authority. During the design phases of the project, if the generation of dust is identified as a major impact, the local community should be included in the communication loop to gain their buy-in at this stage of the project and to make them aware of the potential for releases to occur. As with dust, each project should be risk-assessed for elements such as noise and recommendations made to minimise potential pollution. This includes specifying working hours, and use of equipment or sound proofing.

The key risks identified through the various stages should be noted down within the columns 'Design issues' and 'Build issues' to reflect the existing knowledge available. As there is greater knowledge provided, particularly for larger projects, these risks may change. A further element that is not generally considered is the additional benefits that can be provided from the use of environmental best practice through the project activities. Whilst cost and timescales are predominantly the key factors, many projects have savings resulting in reduced waste disposal to landfill or improved energy and water efficiency. It is recommended for the major risks identified capturing performance measures to equate the benefits from the project.

Clients and tenants are increasingly seeking refurbished properties that provide comfortable working environments, are cheaper to run than their predecessors and provide a more flexible environment to cope with changing work patterns. In addition, a range of sustainable features in their refurbishment projects, which include the use of more 'environmentally-friendly' products in the design phase, are also being sought. There is considerable scope to use products with recycled content in the installation of partitions and building elements that are associated with dividing up the space.

Research undertaken by Davis Langdon shows that medium-sized office refurbishment projects offer good opportunities for product substitution. This particularly applies to refurbishments that can be classed as 'office fit-outs', involving the renewal and upgrading of internal fittings, finishes and engineering services, along with some minor structural alteration.[20] Major refurbishment works that incorporate substantial structural alterations to create an entirely new layout also offer significant opportunities, particularly in changes to the external cladding and structural elements of upper floors.

There is currently very little information available to constructors on how refurbishments can be carried out in such a way as to deliver reduced carbon emissions, e.g. with better designed heating and ventilation equipment or using low/no carbon materials, such as self-ventilating concrete blocking. Also, builders and refurbishers should be encouraged to consider whether they can substitute primary products with secondary or recycled materials, thereby giving environmental gains to the community.

3.5.3   Cradle-to-cradle products

The trend over the past 20 years has been one of changing the office on a regular basis, including the furniture and layout, as technology, working styles and fashion dictated – commonly every few years. Globally the office furniture market is worth

---

[20] *'Why specify recycled in office refurbishment projects'* – for WRAP – http://www.wrap. org.uk

£15 billion. Much of this furniture replaces materials which are often sent to landfill sites. A number of products claim to be from recycled sources, particularly chairs, carpets and desks, however that is at the front end. The challenge is to recycle furniture when no longer required, rather than sending it off to landfill. The image is still of a skip filled with furniture during a refurbishment exercise.

Many modern products are designed and manufactured for single use, with the only end use as landfill, or expensive recycling. Recent examples have included toner cartridges with chips limiting recycling, laptop computers and flat screen televisions as complex electronic equipment not designed for easy dismantling. Many of the products we use include chemicals and finishes which further limit recycling and re-use, particularly furniture and carpets. Some suppliers have take back schemes for items no longer required with charity based companies also willing to take furniture for re-use. A critical problem is that furniture is not designed for quick disassembling or re-use. Companies like Herman Miller and Interface are looking at cradle-to-cradle processes and the construction of furniture to enable quick disassembling and the re-use of materials providing a longer life.

The cradle-to-cradle philosophy is based upon employing the intelligence of natural systems – the effectiveness of nutrient cycling and the abundance of the sun's energy – to create products, industrial systems even buildings which can be used, recycled, and used again without losing any material quality as a continuous cycle.[21] The vision is to create materials, products, and manufacturing systems that are not simply sustainable, but yield sustaining growth in economic prosperity, ecological intelligence, and social value.

Whilst the philosophy may sound unachievable, there are many products appearing on the marketplace which meet many of the criteria and aims of the cradle-to-cradle approach. Regularly used products including carpets, chairs and desks are available at competitive prices with the ability to be re-used and recycled without losing material quality.

The office of the World Resources Institute, Washington DC, sourced cabinets made from biofibre composite materials instead of conventional plastic laminate, besides office doors made from agricultural straw waste. Their office worktops are made from compressed sunflower seeds or Environ, made from soybeans and recycled newspaper – it can be worked like wood, but is 1.5 times harder than oak. The 'linoleum' flooring is made of linseed oil, wood flour, cork flour, natural resins, and a biodegradable jute backing.

Herman Miller has developed bio-based fibres made from renewable resources for use in products in a drive to move away from synthetics. These fibres are derived from non-food-grade corn and potentially other starch containing agricultural plant materials and waste products for use in commercial fabrics, textile backings and carpet applications as well as garments, packaging materials and household products.

---

[21] McDonough Braungart Design Chemistry – http://www.mbdc.com/

Steelcase have developed a chair which is up to 99% recyclable, and composed of up to 44% recycled material, by weight. The production process uses water based glue, water based urethane foam in the seat cushion, PET (recycled plastic bottles) in the back cushion, powder coat paint, as well as recycled steel, aluminium and plastic. Component suppliers use re-usable totes to eliminate disposable packaging where possible. Local suppliers are used where possible, and several manufacturing sites around the world are used to reduce transportation energy consumption. The product can be shipped either blanket-wrapped or in an easy to assemble sized box to reduce packaging and optimise trailer carrying capacity. To help extend its useful life, the chair is easy to reconfigure or upgrade. Seat and back cushions can be replaced. Arms can be added, removed or changed. A headrest and lumbar support can be added. In terms of re-use, the chair can be easily broken down – less than five minutes using common hand tools – making it simpler and easier to get the individual parts into their proper recycling channels.

Another aspect relates to the sourcing and procurement of materials such as timber from sustainable sources. The Forest Stewardship Council (FSC) and PEFC Council (Programme for the Endorsement of Forest Certification schemes) are independent, non-profit, non-governmental organisations with the aim of promoting responsible management of the world's forests. They accredit independent third party organisations who can certify forest managers and forest product producers to trademark standards, providing international recognition for consumers worldwide.

In the United States the energy-labelling scheme, the energy star, covers a wide range of products comparable to the European Union. Homes can qualify under the scheme and are independently verified to be at least 30% more energy efficient than homes built to the 1993 national model energy code or 15% more efficient than the state energy code, whichever is more rigorous. These savings are based on heating, cooling and hot water energy use.

4 Operation of the facility

The operation of a facility carries with it by far the greatest impacts upon the environment, society and the economy of those working within and maintaining the facility itself. Given the time and effort spent in developing the design concept, installation of equipment and procurement of consumable items, it seems surprising that budget forecasting is often significantly different between the anticipated and actual performance of the facility in sustainability terms.

This chapter looks at the operational period in the facility life cycle and some means to be able to derive best practice from the building by reducing impacts and increasing benefits. The changing use of a building in keeping with changes in personnel and equipment means benchmarking and performance assessments should be a regular agenda item and part of the annual programme – not a one-off point in time exercise.

The chapter is split into three sections covering the three main aspects related to the sustainable operation of a facility. The first section looks at the technical aspects of the building, predominantly the energy efficiency and indoor environmental quality and how benefits and improvements can be gained. The second section reviews the non-technical aspects covering waste and resource efficiencies, green transport, pollution mitigation and biodiversity action plans. The final section covers occupant satisfaction to identify and provide facilities to meet demands and improve productivity.

This chapter should be read in conjunction with the management section covered in Chapter 2 in the development of business cases, procurement, management systems and reporting. Elements of performance measurement and management are covered for each of the relevant sustainability categories.

4.1 Maintenance

The technical management of a facility involving the provision of heating and cooling, lighting and on-going maintenance has a significant impact upon the performance of the facility from an environmental and social perspective. The common focus is upon the energy consumption, however many other aspects relating to the use of hazardous materials, occupant satisfaction, life cycle procurement and local community involvement are also critical areas.

This section does not look to duplicate the many publications which exists on the technical operation and efficiency of the equipment, but to highlight the key points which should be drawn out into day-to-day practice, including their implications not only on energy, but staff welfare, materials and life cycle costing.

Section learning guide

This section describes the issues affecting the maintenance of a facility, the various activities involved from a sustainability perspective and means to identify, resolve and promote environmental and social measures:

❑ Indoor air quality and sick building syndrome;
❑ Energy efficiency;
❑ Planned preventative maintenance schedules; and
❑ Replacement programme for capital equipment.

Key messages include:

❑ Efficiency measures should be reviewed as a constant process not as a one-off;
❑ Employ qualified and knowledgeable maintenance contractors;
❑ Fit the correct replacement parts such as filters;
❑ Avoid modification of plant without consideration of the total effect;
❑ Listen to staff feedback and respond to the consensus;
❑ Review the system every year and whenever occupancy changes; and
❑ Poor maintenance can affect energy consumption, productivity and staff welfare.

4.1.1 Introduction

The technical operation of a facility is a complex mixture of reactive task management to meet customer demands, and the proactive maintenance regime to provide a clean and healthy working space. There are a number of tools which can be used to review a facility's current performance and identify areas for improvement. Figure 4.1 provides an energy management matrix covering six main categories to implement an energy technical operation. The categories described are based upon developing and communicating a policy, and achieving it through education, technical implementation and data analysis. The matrix is scored from one as the lowest to five as the highest. The matrix helps to identify where the main gaps exist and where attention should be paid.

Benchmarking can be used to optimise facility performance and enable a comparison either internally or externally with other companies in the same and other sectors producing ratings for the cost of maintenance in cost/m^2 per annum and energy costs in cost/m^2. The benchmarking process can identify areas to:

❑ Reduce costs;
❑ Improve efficiency;

Level	1	2	3	4	5
Energy policy	No explicit policy	An unwritten set of guidelines	Unadopted energy policy set by energy manager or senior department manager	Formal energy policy but no active commitment from top management	Energy policy, action plan and regular review have commitment of top management as part of overall strategy
Organising	No formal delegation of responsibility for energy consumption	Energy management a part time role at a junior level	Energy manager reports to ad hoc committee, but line management unclear	Energy manager accountable to a committee representing users, chaired by a Senior manager	Energy management fully integrated into management structure with clear delegation of responsibilities
Motivation	No contact with users	Informal contacts between engineers and a few users	Contact with major users through ad hoc committee chaired by senior manager	Energy committee used as main channel combined with direct contact with main users	Formal and informal channels of communication regularly exploited by energy manager and energy staff
Information systems	No information system or accounting for energy consumption	Cost reporting based on invoice data, compiled for technical department use	Monitoring and targeting reports based on supply meters Some involvement in budget setting	M&T reports based on sub-metering, but savings not reported effectively to users	Comprehensive system sets targets, monitors consumption, identifies trends, quantifies savings and tracks budget
Marketing	No promotion of energy efficiency	Informal contacts used to promote energy efficiency	Some ad hoc staff awareness training	Programme of staff awareness and regular publicity campaigns	Marketing the value of energy efficiency, performance and benefits within and outside the organisation
Investment	No investment in increasing energy efficiency	Only low-cost measures taken	Investment using short-term payback criteria	Same payback as used for other criteria	Positive discrimination of green schemes, with investment for refurbished and new facilities

Figure 4.1 Energy management matrix. (Source: adapted from organisational aspects of energy management, General Information Report GIR 12 (Action Energy) 1993 and Energy Management Priorities – a self-assessment tool, Good Practice Guide GPG 306 (Action Energy) 2001.)

❏ Improve service; and
❏ Save energy.

A one-size-fits-all approach to maintenance service provision is inappropriate for front of house, and particularly the retail sector, where image is critical. Matching the maintenance programme to seasonal variations in visitor numbers and footfall will help maintain a professional image and meet demand. This means upgrading and refurbishing the space prior to peak seasonal periods to deliver a responsive front of house and retail environment.

Mechanical and electrical

❏ Develop performance indicators covering areas such as planned maintenance quality, energy efficiency, staff skills and competencies and overall customer satisfaction;
❏ Reset temperature set-point for air-conditioned IT/communications rooms to around 25°C rather than a lower temperature;
❏ Ensure installation of sufficient insulation to improve the thermal envelope of the building;
❏ Consider combined heat and power units as a boiler replacement option; and
❏ Survey the building as part of a simple energy audit.

4.1.2   Indoor air quality

Historically, the provision of good indoor air quality (IAQ) within facilities has been limited. Whilst the benefits associated with sustainable buildings, especially energy, waste and water conservation have been recognised, the equally signifi-cant financial and health benefits associated with good indoor air quality are often ignored.

Results of hundreds of studies and reports have demonstrated a significant and causal correlation between improving the indoor environment and gains in pro-ductivity and health.[1] Buildings over the past few decades have been designed and built without a clear understanding of how the indoor environment impacts worker productivity and health, and as a result often create an environment that inhibits productivity rather than enhances it.

Sick building syndrome in offices and workplaces was brought to public atten-tion in the 1970s and 1980s. VOCs from carpets, fabrics and finishes; inadequate air circulation, poor lighting, mould build-up, day lighting and views of the out-doors, noise control and temperature variances – all were contributing to nausea, respiratory problems, skin rashes, lethargy, headaches, and numerous other health concerns. Concern over sick building syndrome led to improvement in building design and maintenance, although 'SBS', as it came to be known, has hardly been conquered.

---

[1] *Indoor Health & Productivity Project* – http://www.ihpcentral.org

---

Case study: Office facility, local authority, Europe – ems

A local authority in Europe was looking to redevelop its offices and increase the amount of usable space it had for occupation. Part of the plan also involved introducing a forced air ventilation system for parts of the office and areas of chambers.

To minimise the space requirements for the plant, much of it was specified on size and, where possible, 'efficiencies' were designed in. One of these efficiencies was that a single, isolated hot water circuit was used to feed the frost/preheat battery on one of the air handling units and the domestic hot water service for part of the premises.

While this did save space and reduce construction/installation costs, it did not deliver any long-term sustainable benefit, either financially or environmentally. The reasons for this were that as the system had been designed and installed as one loop, either the preheat battery and hot water services were both on, or conversely, both off. Although this did not cause a problem in winter, in summer with the system on they had hot water but also needed constant full chilling on the air handling unit to offset the pre-heat battery, or the system could be turned off reducing the cooling requirements but, of course, providing no hot water through part of the domestic services.

Owners of buildings have been held liable for poor indoor air quality, highlighting the importance of managing the risks through operation. With recent mould-related claims, insurance companies in the United States have begun to take defensive action with mould exclusion clauses and premium rate hikes. On the other hand, some insurance companies are willing to offer lower insurance premiums for buildings and facilities with positive environmental effects.[2]

Energy professionals are in a strong position to affect thermal conditions and lighting, while they are often less knowledgeable about indoor pollutants. As a result, to achieve energy efficiency goals, ventilation rates are reduced to the detriment of the indoor air quality and the building occupants breathing that air, thus supporting the misconception that providing good indoor air quality and energy conservation are competing goals.

Studies have also demonstrated how certain features of sustainable buildings have a positive impact on health and well-being, and lead to lower absenteeism. Absenteeism among 3720 employees in various offices at a large US east coast company was 35% lower in offices with higher ventilation rates. Annual cost savings per 100 employees of almost $25 000 could be achieved through a one-time investment

---

[2] Mills, E. 2003b. 'The Insurance and Risk Management Industries: New Players in the Delivery of Energy Efficient Products and Services' Energy Policy. Lawrence Berkeley National Laboratory Report Number LBNL-43642.

of $8000 in improved ventilation systems.[3] A study of 11 000 workers in the Netherlands found that absenteeism due to sick building syndrome is likely to be 34% lower when workers have control over their own thermal conditions.[4] Complaints from building occupants lead to other financial impacts because building maintenance engineers spend unnecessary labour hours. It is estimated that simple efforts to increase comfort could result in a 12% decrease in labour costs attributed to responding to complaints.[5] Less time dealing with complaints leads to more time to complete preventative maintenance, better equipment longevity, and lower operating costs overall.

The first step if complaints are arising is to review the basics:

❑ Ventilation system (CO_2 levels used to measure effective ventilation);
❑ Air quality (VOC levels from materials, particularly after a refurbishment); and
❑ Thermal environment (temperature and humidity levels and variances).

From this initial review, if a specific problem is not identified directly, a building occupant survey should be performed to identify and provide a focus for further investigations. Common causes of poor air quality include the cleaning of ventilation systems, to remove the build up of dirt and bacteria on a regular basis. Cleaning out of the system on a six monthly basis may help to reduce the level of complaints. The erection and removal of partitions in office spaces can also significantly change the flow of air and cause 'dead-areas' where no air circulates. Where the air intake is located is also important to ensure 'dirty' air is not taken up, particularly from vehicle exhaust emissions and cigarette smoke.

4.1.3   Energy efficiency programme

With rising global energy prices, the drivers for reducing energy consumption are becoming more efficient, are self-evident. There are four main activities involved:

❑ Education of staff;
❑ Provision of a monitoring and targeting system;
❑ Reduction of energy consumption through an audit programme; and
❑ Installation of renewable technologies.

These options should be pursued prior to consideration being given to the procurement of 'green' energy or carbon sequestration through the offsetting of carbon dioxide level by buying trees. The key areas to focus upon in a building include

---

[3] Milton, D.K., Glencross, P.M., and Walters, M.W. 2000. 'Risk of sick leave associated with outdoor air supply rate, humidification and occupant complaints', *Indoor Air*, 10 (4): 212–221.

[4] Preller, L.T., Zweers, T., Brunekreef, B., amd Bolej, J.S.M. 1990. 'Sick leave due to work related health complaints among office workers in the Netherlands', *Indoor Air*, 1990. Toronto Volume 1 : 227–230.

[5] Federspiel, C. 2000. 'Costs of Responding to Complaints', *Indoor Air Quality Handbook*. Spengler, J.D., Sarnet, J.M., and McCarthy, J.F. Eds New York: McGraw Hill.

the lighting, heating and cooling systems and controls which consume or manage the largest proportion of energy. See the checklist later in this section for the main data to collect.

The section below does not go into detail on the specific efficiencies which can be derived from lighting, controls, heating and cooling systems, which are covered in more detail within a number of guides provided by the Chartered Institution of Building Services Engineers (CIBSE) and the American Society of Heating, Refrigeration and Air Conditioning Engineers (ASHRAE) and a number of other organisations. The section does provide the methodology to follow, and provides information on the training and information required to make informed decisions on the equipment used and energy consumed.

The benefits which can be provided include:

❑ Peace of mind about energy use and cost;
❑ Reducing energy wastage through staff awareness;
❑ Identifying actions equating to a minimum of 10% reduction in annual energy bills;
❑ Pinpointing opportunities for reducing the amount of energy consumed;
❑ Assisting with better prioritisation of energy related workload; and
❑ Enabling a constant flow of energy reduction measures to be planned.

The first step is to develop an energy policy to inform the company and any other stakeholders about the intention to reduce energy consumption. Targets and deadlines can be mentioned, but details of how these are to be achieved should not be included in an energy policy. The basis for the development of a policy has been discussed in Section 2.3.6.

Performing an energy audit

An audit should be performed over two phases – a preliminary audit to capture consumption patterns and trends highlighting key energy improvement proposals and identifying areas for further investigation. The recommendations will focus attention on actions for the next phase – a detailed site (comprehensive) audit to follow-up agreed actions.

Preliminary audit The survey reviews all significant energy use on the site and is the first step to cutting energy costs and reducing environmental and social impacts. A preliminary audit will deal primarily with energy consumption data (bills and/or meter readings) and does not involve a detailed, intrusive site survey covering building fabric, services, controls systems or performance although these aspects should be considered in any finding.

The main steps are:

(1) collecting data;
(2) analysis;
(3) presentation; and
(4) indicating actions.

(1) Establish quantity and cost of each energy type. Quantity split by fuel type and cost given both as capital cost and carbon level. Source of data will be from meter readings where possible, rather than energy bills and should cover the past three years of operation;

(2) Prepare time based energy consumption graph to show patterns in energy use and identify any anomalies or unusual readings which would warrant further investigation;

(3) Establish the normalised performance index (NPI) value for the building. This allows the relative performance of similar use buildings to be compared and performance assessed against a benchmark value. The NPI is a corrected energy value which takes into account weather variation, building size and variations in operation and can provide useful guidance in setting actions and priorities for further investigation;

(4) Where appropriate data is available, a plot of space heating against degree days can be generated. Assessment of scatter will indicate the level of system control and comparing the regression line with previous years can provide further data on energy trends;

(5) A visual inspection should be made of the facility noting the age and relative condition of the main energy conversion plant e.g. boilers, chillers, air handling plant, etc.), insulation standards and observations regarding the building fabric;

(6) Facility operating personnel will be interviewed to determine any past or ongoing problems, gauge the level of maintenance activity, the level of awareness and understanding of energy related issues and to what level it is incorporated into existing techniques and procedures. Reports on boiler tests where available will be reviewed to determine performance and recommendations made for energy efficiency improvements as appropriate;

(7) External benchmarks to compare the energy performance of different buildings against national averages can be made. The benefit of using this data is that it can compare a building's performance against typical values and also best practice values. When more building types are surveyed and added to the M&T (Monitoring and Targeting) system, a league table can be produced and published (intranet or energy bulletin etc.) which can be used as a motivational tool in the awareness campaign; and

(8) A report should be provided, identifying areas for action and potential cost saving strategies with indicative cost, payback period and expected carbon saving.

Comprehensive audit

Comprehensive surveys normally include measurement of principal energy flows and performance assessment of major plant. The data gathered under the preliminary or concise audit do not allow areas of specific energy consumption and performance to be targeted and often require the installation of sub-meters in order to accurately discriminate between the various energy flows. It may also be necessary to undertake computer modelling of the building to determine how it should be performing.

(1) The comprehensive energy audit and survey will be performed in line with CIBSE Guide F;[6]

(2) Commentary may also be provided on the management culture of the organisation with resultant recommendations on performance improvement measures;

(3) Recommendations will be costed with expected savings in energy use and payback period advised. This may include a scheme proposal for works and anticipated implementation plan. Feasibility should confirm technical considerations, regulatory compliance issues, possible impact on existing operations and list any uncertainties. Reference should be made to future anticipated regulatory changes; and

(4) An action plan for further investigation to verify that benefits have been achieved from the implementation of any recommendations will be presented.

Implementing the initiatives Following the audits, a series of energy efficiency initiatives will be identified. A suitable tracking document should be developed to capture the details and justification, with capital costs (if any), energy savings and paybacks forming part of the standard format (Figure 4.2). The aim of this document is to collate together the projects into a format to describe the overall benefits and installation costs, as well as identifying trends.

A typical process to implement and manage the initiatives includes:

(1) Identify possible initiatives;
(2) Gather relevant energy consumption data;
(3) Present business case for implementation of initiative;
(4) Enter initiatives and related benefits on tracking plan;
(5) Implement initiative;
(6) Confirm energy savings after suitable time period;
(7) Adjust tracking plan where appropriate; and
(8) Use tracking plan for company reporting.

Once no- and low-cost energy saving opportunities have been implemented, monitored and adjustments made to targets, the capital investment projects can be reviewed. Projects that are identified in energy surveys will have to be re-evaluated because any cost saving and payback calculations will have been based on historical data and these calculations will not reflect the lower energy consumption. Capital investment projects can be funded from an internal budget allocation, external funding and grants, or from savings generated from the no- and low-cost initiatives.

Development and implementation of a monitoring and targeting strategy All companies should have a procedure for gathering and storing energy related data. The type of system required for only one building will be different from systems that are used in multi-site scenarios. The development of an appropriate scope to provide an M&T system is given below.

---

[6] CIBSE Guide F, Energy Efficiency in Buildings, Section 18 – Energy Audits and Surveys (ISBN: 1903287340) – http://www.cibse.org

Projects

Proposed project	Est. cost (£)	Est. annual savings (£)	CO$_2$ (tonnes)	Payback period (years)	Comment
Completed projects					
Current projects					

Proposed project	Location	Est. cost (£)	Est. savings (£)	CO$_2$ (tonnes)	Payback period (years)	Project manager	Date work ordered	Comment
1								
2								
3								
4								

Figure 4.2 Energy implementation savings programme.

(1) Data input

❏ Should accept manual and electronic inputs;
❏ Should accept:

  – Invoice data;
  – Meter and sub-meter readings;
  – Half hourly data from suppliers.

❏ Should include data validation and accuracy checks on data being inputted;
❏ Should allow data capture from a building management system.

(2) Analysis

❏ Should allow routine analysis (daily, weekly, monthly, etc.);
❏ Should allow investigative analysis of exceptional consumption;
❏ Should provide:

  – Regression analysis
  – Historical analysis
  – Trend logging and CUSUM analysis
  – Normalised performance indicators
  – Comparisons with benchmarks
  – Weather correction (using site monitored heating/cooling degree days)
  – Comparisons of invoice data versus metered data versus half hourly data
  – Comparison of day and night consumption.

❏ Should promote development of energy cost centres;
❏ Should allow comparison between multiple sites;
❏ Should develop targets for future consumption;
❏ Should support annual tariff tendering procedures.

(3) Reporting

❏ Regular reporting;
❏ Reporting by exception;
❏ User defined reports;
❏ Reporting should be reader specific, e.g. senior management or building operation staff;
❏ Reports should be highly graphic.

The use of the M&T system will provide knowledge on the consumption profiles, trends and operational performance of the facility. It will also help to identify when exceptions in the historical day-to-day operations occur supporting the reduction in energy consumption. Using the energy management matrix described in Figure 4.1, the data should be used not only to review performance, but to communicate effectively with the business, motivate staff and identify where investment will yield efficiencies.

Training programme

A training programme will create an awareness amongst the staff about the energy policy and how this impacts on their jobs, as well as environmental and energy-related legislation. The purpose of training staff will be to ensure they are aware of the energy initiatives to be implemented and the reasons why. Taking the time to explain changes to the internal (and sometimes external) environment will increase acceptance of the changes. Buy-in from senior management will be essential to gain the support from staff. Many of the no- and low-cost measures that will be generated from site audits will require a culture change, and a reason why they have been asked to alter their working practices.

Three training sessions are described covering all parts of the business, since each group will view energy efficiency from different angles and so different drivers and language will be needed to maximise the participation.

❏ Senior managers – Describe the importance of energy efficiency in the workplace; financial aspects related to efficiency savings; tips to reduce energy waste; and full support to the energy management programme and their appointed energy champions;
❏ Energy champions – Focus on information to encourage other staff to change their habits; and
❏ General staff – Cover the energy policy, role of energy champion and how their actions can have a positive impact on the company. This session is important to gain support from all staff and can help to generate new ideas and initiatives. The session will also cover new starters through either part of induction training or a stand-alone session for new employees.

Energy efficiency checklist

Management

❏ Is there an energy policy?
❏ Is there an energy manager and/or a responsible director?
❏ Is there an energy management structure and/or process?
❏ Are consumption records kept?
❏ Are energy reports produced?
❏ Is half-hour data obtained and analysed?
❏ Are sub-meters installed?
❏ Is base load checked and minimised?
❏ Are energy checks and meter readings made as part of the PPM (planned preventative maintenance)?

Lighting

❏ What is the principal type?
❏ What are the lighting (lux) levels?
❏ Is local or individual switching provided?
❏ Is zone switching in place?
❏ Is lighting linked to BMS (building management system)/timer/light sensor?

Heating / hot water

❏ Type?
❏ Fuel?
❏ Temperature control parameters?
❏ Local temperature measurement?
❏ Local temperature control?
❏ Optimiser control?
❏ Is there a summer duty boiler?
❏ Are there local water heaters?
❏ Temperature of stored water?
❏ Age of equipment?

Cooling / HVAC

❏ Type of system?
❏ Air or air/water?
❏ Principal fuel?
❏ Temperature control parameters?
❏ Central or local plant?
❏ Local temperature measurement?
❏ Local temperature control?
❏ Humidification and enthalpy optimisation?
❏ Is cooling managed through a BMS?
❏ What is the age of equipment?
❏ Does the building have opening windows?

General

❏ Any blinds or solar film?
❏ Any heat reclaim schemes?
❏ Any on-site generation?
❏ Any standard design parameters?
❏ Is there a boiler shutdown programme?
❏ Are variable speed drives to fans and pumps installed?
❏ Does BMS shutdown or loadshed?
❏ Is free cooling available?

4.1.4 Planned maintenance schedules

Regular maintenance and inspection of equipment ensures they function effectively and operations are optimised, reducing the consumption of energy and the potential to break down. The use of a condition based maintenance variable against equipment types will maximise the use of the equipment.

Case study: Office facility, London, UK – ems

Following major redevelopment, an office building in London had the cold water storage tank built on trestles and a horizontal calorifier placed under the tank in between the trestles as a 'space saving' measure. The issues here were that because the space available was already limited, the calorifier could not be located without having its insulation jacket removed and the use of trestles meant that the top of the tank itself was a mere 10 cm below the plant room ceiling.

While the design and installation went well, and both systems were successfully commissioned, it was only following an environmental audit that the ongoing problems they had created were identified. Theses were:

❏ neither the cold water storage tank nor the calorifier were accessible for inspection;
❏ no routine maintenance could be performed on the tank (ball valve replacement etc.) as it was not accessible;
❏ the fact that the calorifier had lost its insulation jacket meant that heat generated from its operation was transferring to the base of the cold water storage tank, heating the contents within and pushing the temperature of stored water to within the *Legionella* hazard band; and
❏ neither system, therefore, actually worked properly.

Case study: Office facility, Birmingham, UK – ems

A building in Birmingham had a cooling tower replaced (after some 15 years) with a new facility. The new cooling tower was specified for '314' grade steel as opposed to the more expensive but better '316' grade. The water treatment regime for the system was also amended with an automatic brominator (oxidising based biocide) installed and biodispersant hand dosed.

Unfortunately, the calculations for the water treatment regime had been based on the old tower system volume rather than the new. This, coupled with the lesser specified steel, caused internal corrosion of the system that resulted in the pond having to be recoated within the first six months of use. Longer term damage meant the tower itself had to be replaced within four years of installation rather than the eight to ten originally cited.

Planned maintenance (PM) schedules can also be used for activities performed on a regular basis such as housekeeping tours and checks to record labour needed and confirmation of compliance. Typical equipment will include interceptors, bunds, spill kits and waste reviews.

A PM schedule should be developed for equipment, with timescales for maintenance and inspection attached individually for each PM.

❑ Identify all equipment on site which requires maintenance;
❑ Identify all equipment which requires statutory inspection and determine the associated frequency of inspection; and
❑ Identify all equipment on site that requires calibration and the frequency at which this is required.

New equipment should be identified and registered as they arrive to ensure the appropriate inspection is carried out. For each piece of equipment identified, the following information should be collected.

❑ Equipment details including serial number and location;
❑ Details of the maintenance work or calibration required;
❑ Frequency of maintenance or date next due;
❑ Details of the approved contractor for the maintenance work; and
❑ Frequency of inspection or date next due.

An assessment of the schedules takes place on a continual basis through on-going discussions to take into account the need to increase or decrease the level of inspections for each scheduled PM.

4.1.5   Ozone depleting substance replacement programme

The movement away from fluorinated compounds has been well publicised from the Montreal Protocol Agreement in 1987 leading to the bans of CFCs, halons and HCFCs amongst other compounds.[7] Moves to ban HFCs will come into effect once a viable and cost-effective alternative has been found. The development of a replacement strategy will provide the flexibility and cost-effective approach to the forthcoming changes, with an effective chiller maintenance programme providing a useable life of over 25 years for the unit.

In order to plan for compliance changes in the future, it is necessary to understand the inventories of the current equipment across the location and their existing status, the co-ordination with asset management plans and equipment surveys. This information is required prior to developing an option evaluation, project implementation and on-going maintenance plans.

(1) Establish the refrigerants being used and level of compliance:

❑ Establish what refrigerants are currently being used;
❑ Ensure leakage procedures are in place and followed;

---

[7] Montreal Protocol – http://www.unep.org/ozone/Montreal-Protocol/

❑ Records of annual leakage checks within refrigeration systems are kept;
❑ Waste disposal records are kept and their compliance with legislation; and
❑ Emergency action plans are present and are being followed.

(2) Identify the ownership category which will depend on the type of equipment owned. There are two key categories:

❑ *Category 1* – Small self-contained units that are mass-produced in a factory using hermetically sealed compressors. For example, domestic fridges and freezers, small retail displays, 'through the wall' air-conditioners. Category 1 systems are usually leak free for the whole of their working life.
❑ *Category 2* – Small, medium and large systems using more complex items of refrigeration equipment and usually requiring some on-site systems assembly and refrigerant filling. Category 2 systems are susceptible to refrigerant leakage and often require regular maintenance. Typical examples range from pub cellar coolers, small cold stores, remote condensing units in grocery shops and split system air-conditioners to much larger plants such as supermarket central systems, industrial systems and water chillers.

(3) Assess the criticality of the refrigeration systems. The HCFC phase-out strategy should be dependent on the criticality of the refrigeration equipment to operational activities. Where the failure of a system may lead to production or customer satisfaction being affected, the phase-out should be given a high priority.

(4) Develop replacement programme:

❑ *Response for category 1 users*. Refrigeration systems in this category are very reliable and often run for up to 20 years without requiring maintenance. Under these circumstances your actions should be to:

(a) Continue running your existing plant until it reaches the end of its useful life;
(b) Make an appropriate contingency plan in case the system breaks down and loses its charge of HCFC; and
(c) Plan how you can safely dispose of an old system without illegally venting refrigerant to the atmosphere.

The nature of an appropriate contingency plan will depend on the type of equipment owned and the strategic importance of the systems. Some typical contingency plans include:

(a) Ensuring the rapid purchase of a new system using an ozone-friendly refrigerant; and
(b) Agreeing with a contractor that if you have a breakdown the system can be repaired and then topped up with the original type of refrigerant (only for HCFCs) or a retrofillable alternative. Category 1 plants contain only a small amount of refrigerant, so the cost need not be excessive even if the price of HCFCs rises considerably.

❏ *Response for category 2 users.* If you are a category 2 user then a carefully structured approach should be adopted. There is a wide range of technical solutions that can be applied to achieve ozone depleting substance phase-out. In some cases the best line of action is quite clear, but in most instances there are competing options that require assessment. It is likely that the most cost-effective solution will involve a combination of several different activities such as purchase of new plant or conversion of some existing plant. It should be noted that the structured programme of activities is particularly important for those users who own a number of refrigeration systems. Owners of a single large plant will generally have fewer options to consider.

Within large organisations it is important that senior level commitment is given to HCFC phase-out and that someone is made responsible for undertaking the steps.

4.2   Operation

The operation of a facility has by far the greatest impact on the environment and society. Given the inclusion of sustainability measures in the purchase of any new space, or projects as described in Chapter 3, the emphasis of this section is to continue the good work through maintaining the performance through the non-technical operation of the facility.

Maintaining operations is not as simple as it sounds. It does not simply mean capturing data, measuring it and continuing to operate the facility based upon the data. If that were the case, much of the FM function could be an automated process. It means keeping the facility aligned with business operations, priorities, working times and staff numbers.

Section learning guide

This section reviews sustainability measures and good practice items for incorporation into day-to-day operations from simple no cost measures through to performance measurement:

❏ Sustainability impacts of FM services;
❏ Inclusion of performance measures to incentivise whole-life value;
❏ Good practice and performance measurement for waste and water consumption;
❏ Development of a green travel plan; and
❏ Pollution control and mitigation.

Key messages include:

❏ Good practice is a culture, not a one-off exercise;
❏ Minimise impacts and promote benefits – where an impact occurs, it can be balanced through incorporating benefits; and
❏ Culture change from the top is critical to the success.

4.2.1 Introduction

There are a number of good practice measures which can be incorporated into facilities management services. A series of examples based upon typical service lines are described below. The various measures may not be applicable to all organisations, but should be used as a prompt and guide.

Catering and vending

❑ Install aerating kitchen taps where feasible;
❑ Source labour locally;
❑ Use permanent/re-usable crockery and cutlery where possible to minimise waste. Where not feasible, procure plastic or paper alternatives that can be recycled easily;
❑ Provide bulk portions of products such as sugar and milk (and provide milk jugs and sugar bowls) in place of individual portions;
❑ Bulk buying from local suppliers helps reduce the amount of packaging generated and reduces transport costs, which helps the local community;
❑ Investigate the opportunities to compost any waste food or look at central composting through the local authority or community group;
❑ For tea and coffee making, whether vending machines or kettles, ensure they are energy efficient and can be recycled at the end of their life;
❑ Purchase energy-efficient equipment in catering areas; and
❑ Cooking oil can be recycled, but again there is a minimum quantity for collection.

Cleaning

❑ Specify environmentally-friendly cleaning products, avoiding disinfectants containing trichlorophenol, cresol, benzalkonium chloride and formaldehyde, and detergents containing phosphates;
❑ Bulk order cleaning products to save on packaging where feasible;
❑ Clean windows regularly to maintain the level of available daylight;
❑ Install basin plugs in all kitchen, cleaning and wash-hand basins to save water; and
❑ Provide centralised recycling facilities.

Transport

❑ Ensure the provision of teleconference facilities to minimise transport required for meetings;
❑ Get fleet and staff vehicles serviced regularly – more than 90% of badly polluting vehicles can be re-tuned within 15 minutes;
❑ Make public transport details available to all staff;
❑ Look at the viability of shuttle buses with local businesses;
❑ Provide changing and shower facilities for cyclists;

❏ Provide lockers or other storage space for cyclists to keep clothes and cycling equipment; and

❏ Install cycle locking points to serve at least 10% of office occupants.

Stationery

❏ Think before you print – do you really need a paper copy?

❏ Print documents on both sides of the paper – making sure you proofread it on the screen first. If draft copies are needed, print them on scrap paper;

❏ Always photocopy using the duplex option when available, and remember to return the setting to one copy if you've done several; and

❏ Set up an office paper recycling scheme. White paper is of high grade and so is in demand from the paper industry. The market for this quality is much more stable than lower grades.

Computers and electrical equipment

❏ When purchasing electronic equipment, stipulate minimised packaging, e.g. on a pallet with shrink wrap, and the sleep mode activated;

❏ Try to repair equipment before having it replaced – invest in a long-term maintenance contract for any appliances;

❏ When buying replacement equipment such as printers, photocopiers and fax machines, choose those with a duplex option, mailbox option and recycle the outdated items;

❏ Recycle computers/electronic equipment at the end of their life;

❏ Mobile phones and their batteries can be recycled;

❏ Set your computers to 'power save' when not at your desk; and

❏ Switch off computer screens and hardware at night and over weekends.

Miscellaneous

❏ Make sure your mailing lists are up-to-date to avoid sending out unnecessary details;

❏ Join your local business in the environment group; this provides a great way to exchange practical ideas with others who are trying to improve their environmental performance;

❏ Only boil enough water in a kettle for yourself to use;

❏ If you have a choice, buy recycled products or goods from sustainable sources such as Traidcraft;

❏ Switch to re-useable and returnable packaging;

❏ Install percussion taps and flow regulators to reduce water consumption; and

❏ Switch off appliances instead of leaving them on standby, e.g. TV, computer screen, which uses up to 20% normal use electricity and reduces the risk of a fire.

The various service lines described in the opening Section 1.1 all have a variety of environmental and social impacts. Table 4.1 provides an overview of these impacts captured against the 14 sustainability categories covered throughout this book. The table describes the level of involvement each of the service lines requires for the sustainability category based upon the number of stars. The greater the number of stars, the higher the level of involvement the service line has.

The table should be used to focus attention on where sustainability benefits can be made, align with service providers (internal or external) and support the development of performance measures or benchmarks. For example, the secretarial function will involve not only local labour and human rights, but also payment on time and treatment of staff.

4.2.2   Green transport

With an increasing number of cars and single occupancy vehicles travelling today, and the trend increasing across the globe, the provision of green transport programmes, particularly for existing overstretched sites is becoming the gap between relocating and remaining on site. Parking spreading onto neighbouring streets which affects the local community is commonplace.

Sites in the United States are placing a greater number of car parking spaces than the number of employees operating in the building to size for any increases in building headcount through efficiencies of working. This provides no incentives for green transport and minimising congestion on the roads.

The provision of a green transport plan has greater impacts than reducing car parking spaces. There are the obvious associated cost savings, but also a greater choice of facilities to select from in the rental market and better neighbourhood relations.

The steps to develop a green transport plan are provided below. At the outset, the objectives for the project – a project brief – should be documented. Typical objectives are provided below:

(1) Maximise the number of spaces available on site to ensure all are utilised efficiently and effectively;
(2) Minimise the number of additional personnel attending site if not required or find alternative parking arrangements;
(3) Use electronic tools to link room booking with parking space booking to raise the profile of parking availability on site;
(4) Incentives to encourage alternative modes of transport and car sharing; and
(5) Utilisation of shuttle bus provision to the station and local environs.

❏ Investigate the use of cycling and the requirements that would bring;
❏ Investigate the use of walking and the requirements that would bring;
❏ Investigate the use of an LPG shuttle bus service and conclude if usage is efficient.

Table 4.1 Sustainability involvement of selected service lines. The stars correlate the sustainability impact between the categories and the service line.

Sustainability category	Building management	Catering	Cleaning	IT and data management	Ground management	Security	Procurement	Project management	Telecomms	Secretarial	Postal	Health and safety
Management	**	**	**	**	**	**	**	**	**	**	**	**
Emissions to air	**	*	**	**	**			***				**
Land contamination	**	**	**	***	***		***	***				*
Workforce occupants	**	**	**	***		***	***	***	**	**	**	***
Local environment and community	**	***	***		***	***	***	***		***	***	***
Life cycle or building/products	***	***	***	*	**		***	***				***
Energy management	***	**	**	***			***	***	**	*		
Emissions to water	***	***	***		**	*	***	***				
Use of resources	**	***	***	***	**		***	***				
Waste management	**	***	***	***	***	***	***	***	***	**	**	
Marketplace	**	***	***	**	***		***	***	**	**	***	***
Human rights	**	***	***	*	***		***	***	*	***	***	***
Biodiversity	*	**	*		***		*	**				
Transport	*	**	**		**	**	**	**		**	**	**

(6) Undertake business unit surveys to investigate how people currently travel to the site.

❏ Investigate car sharing, use of smart cars, etc.

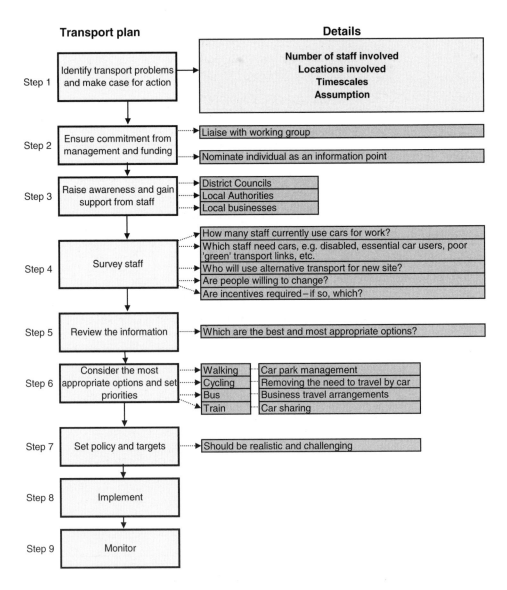

The objectives will include forecasting future car parking supply and demand for future population levels and implementing management strategies that enable these populations to be accommodated.

Identify potential parking shortfall

The first task for the study will be to identify the likely future potential parking shortfall. This is an important task as it will determine the number and type of

Table 4.2   Parking requirements based upon type of worker

Fixed workers	Workers that will be based on site 100% of their time. Fixed workers will always require a parking space if they travel by car.
Flexible workers	Workers who are not based on site 100% of their time, but may visit the location 2, 3 or 4 days per week and will spend the whole day at the location. They will require a parking space for the whole day when on site if they travel by car.
Mobile workers	Workers who may travel 2–3 days per week but for meetings only. They may only be on site for a few hours therefore would not require a parking space for the whole day.

initiatives that will need to be introduced to encourage staff, contractors and visitors to use alternative modes to single occupancy car travel. Potential parking shortfall will be predicted by reviewing existing car park usage and comparing this to average daily attendance on site. The resulting ratio can be applied to the potential site population to give estimates of future parking demand. This future demand can then be compared to the existing parking supply to identify the potential parking shortfall.

Part of this exercise will be based upon the type of workers within the facility, hours of work and the calculation of the maximum and typical number of people on site. Table 4.2 provides a summary of the three main types of worker and the parking requirements.

Identify existing travel habits and how these might change in the future

Having identified the potential parking shortfall it will be necessary to identify how existing staff, contractors and visitors travel to and from the site and how these habits might change in the future if various travel initiatives are improved/introduced. This information will be identified by undertaking a site-wide travel survey, including FM staff and contractors who tend to live locally and use alternative modes of transport. A typical travel plan is detailed later in this section.

Develop a map of staff home locations

The survey will be supplemented with a plot and analysis of staff home postcodes to identify where staff live. This will help identify the most appropriate initiatives to implement to promote the use of alternative modes to single occupancy car travel.

Prepare a travel plan

Having identified future potential parking shortfall, analysed the results of the travel survey and identified relevant initiatives, it will be possible to identify relevant initiatives to encourage site occupants to use alternative modes of travel to single occupancy car travel. A travel plan will be prepared accordingly that will set out the objectives and background to the travel plan, the initiatives to be introduced and the targets that need to be met to avert a shortfall in parking provision.

Implementation of travel plan

The type and number of initiatives that will need to be implemented will be determined by the predicted shortfall in parking provision and the results of the travel survey. Initiatives that may be implemented include:

Public bus use It will be necessary to identify any existing/potential use of services from the travel survey. If there appears to be potential to use the service, then a discount on season tickets could be obtained. The potential to use liquified petroleum gas (LPG) should be investigated including the potential to secure any grants that may be available.

Car share intiative 'Off-the-shelf' car share database software packages are available, with the success of car share schemes dependent on the provision of various initiatives. These include the provision of:

❏ Priority car share parking spaces located in the most accessible locations;
❏ Guaranteed rides home in emergencies; and
❏ Monthly prize draws.

Cycling / walking / running There are numerous initiatives that can be introduced to encourage staff to cycle, walk or run to and from the site. These include the provision of cycle/footways to and from the site, safe and secure cycle storage facilities, showers and changing facilities, shower packs, cycle maps, cycle maintenance equipment. A monthly bicycle user group could be set up to enable cyclists to discuss relevant issues and cycle training could be provided to resolve any Health and Safety Executive concerns that may arise from encouraging cycling.

Park and ride A park and ride scheme could operate by identifying an (or a series of) off-site car park(s) and operating a dedicated shuttle bus between the car park(s) and the site. It would be necessary to identify an existing off-site car park that can be legally used for this purpose.

Marketing

A key requirement to the successful implementation of a travel plan is publicity and marketing. A uniform branding and a catchy strap line will raise the profile of the plan.

Further initiatives

Depending on the predicted parking shortfall it may be necessary to implement further initiatives to encourage use of alternative modes:

❑ SMART car initiatives – three SMART cars can fit into one standard parking space. This is an innovative initiative although it would be necessary to seek legal opinion to identify if such an initiative could be implemented legally. It would also be necessary to offer priority parking spaces to SMART car users;

❑ Personalised travel planning – this is the provision of individually (bespoke) tailored information on travel alternatives for particular journeys or more general travel patterns;

❑ Demand responsive transport – an initiative that allows staff to book a shuttle bus on-line to pick them up/drop them off at a specified time/date; and

❑ Parking cash out – this is an employee benefit that offers staff the opportunity to forgo their employer provided car parking space in exchange for a financial reward.

Ongoing operation

To ensure the travel plan is working both effectively and efficiently it will be necessary to closely monitor the ongoing modal split through further travel surveys, recording car park usage and shuttle bus patronage, and regularly reviewing future staff relocations to and from the site. Travel plans usually have an identifiable travel co-ordinator, with a hands-on role in pushing forward initiatives and ensuring that they run effectively. A representative should be identified with responsibility to liaise with the local community, businesses and travel networks to provide discounts and shared practices where possible.

Travel plans rely on the backing of senior management – the more visible the better. It is unreasonable to expect travel co-ordinators to implement contentious measures without a clear mandate from those running the organisation.

To support the green travel plan, the use of electronic tools may help manage the allocation and measure the usage of parking spaces:

❑ Centralised desk and parking space booking – central co-ordination of desk and parking space booking would ensure all staff visiting the location have a parking space;

❑ Centralised room booking – a desk and parking space booking system linked to be able to co-ordinate together; and

❑ Visitor management – a visitor management tool linked to the desk and parking space system to manage the known visitors arriving on site and to ensure sufficient spaces are available.

TRAVEL PLAN

Q1. Is the Site your normal place of work? (Yes/ No)

Q2. How many times a week do you travel to the Site? (From 1 to 7)

Q3. What is your main mode (by distance) of transport? (please select *one* from the list below)

☐ Car Driver		☐ Motorcycle
☐ Car Passenger		☐ Walk
☐ Bus		☐ Bicycle
☐ Train		☐ Other (please state) ....

Q4. If you used another mode of transport, which mode did you use? (Choose from list above)

Q5. Where did you travel from – please enter town and postcode? ....

Q6. What time do you normally arrive at work?   ....

Q7. What time do you normally finish work?   ....

Q8. Do you travel on work business during the day? If yes, where do you travel? ....
What is your main mode (by distance) of transport?

☐ Car Driver		☐ Walk
☐ Car Passenger		☐ Bicycle
☐ Bus		☐ Other (please state) ....
☐ Train		
☐ Motorcycle		

TO BE COMPLETED BY CAR DRIVERS AND PASSENGERS ONLY.

PROCEED TO QUESTION 12 IF YOU TRAVEL BY AN ALTERNATIVE MODE.

Q9. Why do you use a car? (select up to three responses that are applicable and rank them 1, 2, 3, where 1 is the most important)

☐	Door to door convenience	☐	Public transport overcrowded
☐	Cheaper than public transport	☐	Public transport unreliable
☐	Quicker than public transport	☐	Public transport not available during hours
☐	Company car available	☐	Dissatisfied with public transport for other reasons
☐	Company parking available	☐	Carrying goods or shopping
☐	Need car for other business	☐	Other (please specify) ....
☐	Need car for other private journey		

Q10. Would you join a car-sharing scheme? (Yes/No)

Q11. What would be your major concerns of (not) joining a car-share scheme?

☐	Prefer the privacy of travelling alone	☐	Do not travel directly from/to work
☐	Hours of work vary too much	☐	Other (please specify) ....
☐	Often travel on business at start or end of day		

POSSIBLE IMPROVEMENTS – TO BE COMPLETED BY EVERYONE

Q12. If you could not / do not use your car to travel to work, what alternative mode of travel would/ do you use? (Please tick one box below)

☐	Bus all the way (go to Q13)	☐	Walk (go to Q16)
☐	Train and shuttle bus (go to Q13)	☐	Could not get to work
☐	Bicycle (go to Q15)	☐	Other (please specify)

Q13. If you answered by bus or train, which service would / do you use? ....
....

Q14. If you answered by bus or train, which of the following measures would you like to see implemented? (select up to three responses that are applicable and rank them 1, 2, 3 where 1 is the most important)

☐	Increased frequency of service	☐	Better quality service
☐	Faster service	☐	Other (please specify)
☐	Reduced fares		

Q15. If you answered by bicycle, which of the following measures would you like to see implemented? (select up to three responses that are applicable and rank them 1, 2, 3, where 1 is the most important)

☐ Showers and changing facilities on site

☐ Introduction of interest-free cycle loans

☐ Secure cycle parking

☐ Provision of bicycles for employees to borrow

☐ Other (please specify)

Q16. If you answered by walking, which of the following measures would you like to see implemented? (select up to three responses that are applicable and rank them 1, 2, 3, where 1 is the most important)

☐ Better footpaths and crossing facilities adjacent to the site

☐ Better footpaths and crossing facilities beyond the site

☐ Other (please specify)

4.2.3   Waste management

Historically, waste management has been primarily the concern of manufacturing and industrial processes which in high volumes of potentially hazardous materials. This was a legislative and cost burden that had to be managed effectively to reduce the impact on an organisation. Within the office environment, waste management was under the control and commonly procured in line with the cleaning provision and therefore hidden from visibility. This section describes the impacts and reasoning of why waste management has now become a topic of conversation and concern, and one whose visibility is required to provide effective management. The role of FM in managing waste including legislative requirements, recycling, awareness programmes and performance measurement is discussed.

The principles of effective waste management – avoid, reduce, re-use, recycle and recover – are well established but seldom implemented in their entirety as part of a well-planned business strategy (Figure 4.3). The increasing impact of environmental legislation – and, in particular, the introduction of the 'producer pays' principle – is now driving companies to focus on this aspect of their operations. The aim is to develop a coherent approach to this area of activities, which can produce many benefits in financial and environmental terms, as well as reducing the risk of non-compliance.

The safety of disposal methods and their implications on public health are also a serious consideration. Landfill sites can produce leachate that, unless properly managed, can contaminate the water supply. Also as a result of breaking down of materials within the site, methane and CO_2 are produced. These are greenhouse gases and can add to global warming. Incinerators give off dioxins that are potentially hazardous to health. Added to this, there is a shortage of available landfill

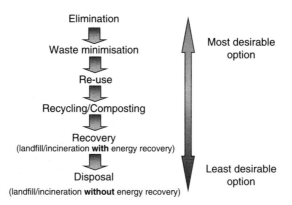

Figure 4.3 Waste hierarchy.

space in some areas of the country and having to meet strict European emission rulings led to many incinerators closing down in 1996. These factors increase the need to reduce the amount of rubbish that is thrown away.

Retailers have already felt the effects of consumer pressure on their business in the wake of public concern about environmental issues. As more local authorities increase and improve their arrangements for domestic waste reduction, recycling becomes a normal part of home life. In addition to these environmental concerns, there are very practical reasons why every organisation, large or small, should look at ways of reducing the amount of waste they produce.

The cost of waste

All businesses must, by law, have proper arrangements for the disposal of their waste. This normally involves paying a waste management company to provide containers for the rubbish and empty them on a regular basis. There is usually a charge for commercial waste collection and so by reducing the amount collected, companies can reduce the cost of waste disposal. In addition, looking at processes and operations within the organisation that are wasteful, where careful use of resources would result in fewer materials needed, will save money on top of the environmental and social implications including traffic, and the location of a waste plant in the local area.

A good way to start to identify areas where there could be greater efficiency is to conduct an audit which considers every aspect of the organisation and the implications for resources and waste on society and the environment. Many of the savings can be achieved with little or no investment from the companies involved.

The following sections detail the key activities which an FM department will be involved in, to effectively manage and provide best practice in the reduction and reporting of waste over and above legal requirements.

Operational waste management and recycling Traditionally FM has extensive experience of managing all types of waste originating from an organisation's operation from normal office waste to the more specialised varieties, e.g. clinical, building waste, oil or chemicals. The waste removal service should be performed by an approved supplier which has a registered waste carrier's licence and is managed either on its own, particularly for most specialised wastes, or in conjunction with the cleaning or grounds maintenance service for general waste or garden waste.

In addition, FM must ensure that the waste service is fully legally compliant against the requirements of duty of care legislation – cradle-to-grave – taking the waste from generation within the organisation through any transfer station or recycling depot to its final end point, be it recycling, landfill or incineration. A template for the capture of the waste details is captured in Table 4.3.

FM is also responsible for ensuring that efficient processes are in place for waste handling, movement and control; the role of FM should allow for integration of waste management and recycling with other FM activities. This enables the adoption of a more proactive approach on behalf of clients – either in-house or contracted.

An example of this is by ensuring that appropriate waste removal arrangements are in place in advance of building or refurbishment activities, thereby avoiding unnecessary disruption and expenditure. Through this process appropriate waste contractors are identified in advance of the move. Unwanted items are either disposed of to pre-labelled bins, or labelled for waste to be removed directly by the waste contractor. This provides maximum potential for recycling and recovering the waste, reduces the potential for waste disposed of incorrectly, or left bagged outside.

Waste flow analysis The operational activities within a building are continually changing, particularly with increasing fluctuations in building occupancy, and changes in activities performed. All of these will lead to changes in the volumes of wastes arising and types of waste streams needing to be managed. The use of the waste flow analysis technique involves reviewing all aspects of a company's waste management activities to ensure that these are conducted in a way which is both cost-effective and environmentally beneficial, as well having minimal impact in terms of building management.

Areas involved in this review may include:

❏ Identification of origins, types and volumes of waste. This should include both the general day-to-day waste arisings and specific arisings which may result from ad hoc activities such as planned maintenance works and clearance of interceptors;
❏ Movement of waste prior to storage and disposal including routing for the cleaners, towards the storage location, availability of space for bins to be located on the office or shop floor and sufficient signage; and
❏ Waste segregation and storage arrangements capable of taking the capacity of waste being generated. If there is excess capacity, removal of this or a reduction in pick-ups per week to reduce costs and improve efficiencies.

Table 4.3 Waste form for collecting legislative information on waste carriers and disposal sites

Waste type	Waste contractor	Transfer note	Expiry	Carrier registration	Expiry	Disposal site	Site licence
General office waste							
Wood							
Plastics							
Aluminium cans							
Cardboard							
Confidential paper							
Glass							
Sanitary							
Food waste							
Cooking oil							
Oil/water							
Grease interceptor							
Toner cartridges							
Refrigerant gas							
Builders' waste							
Garden waste							
Chemical waste							
UPS batteries							
Fluorescent tubes							
Soluble/mineral oil							
Asbestos							
Cutting fluids							

Waste costs – manufacturing organisation

An organisation involved in the 'just-in time' assembly process had performed a waste management review of their performance in the early 2000s to identify low recycling rates and a high cost base. The service was tendered and was awarded to a local company. Two key performance indicators (KPIs) were set – the percentage of waste recycled, and the cost of waste disposed of. Over the next two years, both indicators improved dramatically, with the overall cost dropping by over 50%. Needless to say, the client was impressed.

A review of the operations yielded some surprising results:

❑ The cost of service only covered the transport and direct disposal charges made. Costs for labour, consumables and equipment were excluded;

❑ Labour charges were for ten staff at more than double the traditional hourly rate, with overtime a regular occurrence – however, only eight staff operated on site;

❑ The waste contractor used the additional labour charges to offset the disposal charges and pay-off recycling rates – therefore the figures provided for the indicators were false;

❑ Over the two years, the true total waste cost including labour, consumables and equipment had increased by more than 10%; and

❑ Savings of 40% in the total cost were identified through rationalisation of labour personnel and charges, stopping unnecessary overtime, increasing recycling and the revenue obtained, and revising the equipment needed.

This type of review should be performed on a regular basis to ensure efficiencies are maintained and a service appropriate to the organisation is achieved.

Waste minimisation and recycling The increasing levels of landfill disposal tax and increased employee and customer awareness have underlined the importance of companies reducing the amount of waste they generate. This is through a process of waste minimisation which involves tracing the origins of the various types of waste and aiming to minimise this at source wherever possible.

A key element of the approach entails a review of waste generated throughout the organisation to gather baseline data of the volumes and costs associated with the waste generated and the processes from which the waste arise. The aim of this is to identify areas that generate a high volume and cost of waste in comparison to the flow-through of product (Figure 4.4).

Waste minimisation should incorporate the following activities:

❑ Liaison with suppliers to reduce the amount of potential waste coming onto the site from external sources such as from excess packaging, excess deliveries or junk mail;

❑ Reviewing operational activities and work processes is also important in order to reduce the use of raw materials and encourage the holding of data electronically, rather than in paper form; and

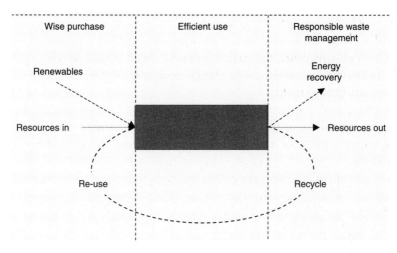

Figure 4.4   Life cycle of material consumption and waste generation.

❏ Review of the procurement process to ensure over deliveries and excess product is not delivered. This is particularly appropriate for high-volume orders based on bulk discounts.

A review should be made of the most appropriate waste minimisation activities that can generate a reduction in waste and also cost savings. A programme should be put in place to develop these plans with timescales and owners for each plan. It is important that any plans put together have the support of senior management to drive through the changes and ensure they are maintained.

Recycling and recovery can make a significant contribution to minimising waste such as through reconditioning to allow items to be re-used, recovery of raw materials, or through heat generation on disposal. The common office environment has the ability to recycle and recover over 70% of all wastes generated through segregation either on or off site. However, some industries and offices struggle to achieve more than 10% recycling which impacts on costs and efficiency.

The recycling of waste can be achieved through on-site segregation, with a variety of bins and coloured bags used to distinguish the various streams. There are a number of bins available for plastic cups (becka bins), aluminium cans, flat packed cardboard, glass bottle banks, and paper bins. Much of this is dependent on storage within a building until a sufficient load is collected to justify a pick-up. The major hurdle in this instance is ensuring waste is segregated effectively by staff with minimum cross-contamination, which is linked to the awareness and training provided.

An alternative is to utilise off-site segregation through a materials recycling facility (MRF) where uncompacted waste is taken and recyclable waste materials are removed. The benefits are that waste does not have to be segregated on site, removing the need to store waste, although without the visibility of segregation taking place, staff need to be re-assured through a different form of awareness that the recycling levels are being achieved.

Education and training Initiatives to reduce and recycle waste are unlikely to be successful without the wholehearted co-operation and involvement of the workforce. Often a major culture change is necessary which is required through various training and communication programmes designed to raise levels of employee awareness and engagement.

Any training and awareness programmes which are developed must be continued for a period of time to push through the changes, with enforcement required where necessary to stop cross-contamination of individual segregated waste streams.

An activity which has worked well is the competition achieved through various departments or floors particularly in an office environment or manufacturing cells in industry. A reportable factor is chosen such as waste volumes or cross-contaminated bins which is checked on a regular basis and publicly reported to all staff through a league table. This has the double incentive of learning and transferring best practice across all locations, but also of pushing poor performers to improve their image.

Waste contract review Commonly waste disposal is handled through a variety of disparate contractors dealing with specific strands of waste often procured as a number of small value contracts. This will include confidential paper through security, general waste through the cleaning contractor, fluorescent tubes and special wastes through another party. This can be exacerbated where more than one site is managed, with a number of contractors providing services autonomously across the portfolio.

A waste contract review identifies the various waste management activities being undertaken and the associated contractual arrangements. This will include the various costs invoiced and volumes generated. It ensures that measures to streamline processes, including increased recycling and recovery of wastes and reduced capacity if required, are incorporated into the service specification and that all waste management activities are consolidated into one contract, which can be offered to a single supplier.

A variety of waste options is available in Table 4.4 which captures the advantages and disadvantages for each of the various options.

Packaging contracts in this way enables organisations to increase the amount of recycling and recovery they undertake at a reduced cost. Organisations can typically achieve savings in the region of 25% through contract consolidation, while doubling levels of recycling and recovery. Having a single point of contact for all waste management issues also produces considerable savings in administration. For larger portfolios, greater direct and indirect savings can be realised through sharing best practice and efficiencies of scale.

Monitoring and data capture FM is commonly responsible for the vast majority of waste streams resulting, and the capture of volumes and costs. For each of the various waste streams it is important to capture the most accurate waste figures available. Commonly for non-hazardous waste streams, a waste contractor will

Table 4.4 Waste disposal options – advantages and disadvantages

Management option	Possible advantages	Possible disadvantages
Waste elimination and minimisation	☐ The ultimate waste reduction strategy is to arrive at zero discharge. Waste reduction should ideally be a key element of an effective waste control strategy. ☐ Provide improved public relations. ☐ Waste minimisation involves reducing pollution to all environmental media – air, water and land. ☐ Reduce environmental liability and insurance costs and yield income from the sale of re-usable waste. ☐ Offset the increasing cost of off-site disposal, on-site pre-treatment costs and water, fuel and raw material costs, which could all be reduced by avoiding the creation of waste in the first place.	☐ Unlikely to be achievable. ☐ Not cost-effective. ☐ Often the obstacles to introducing waste minimisation schemes are institutional rather than financial or technical. ☐ Payback periods must be fully examined prior to commencing a waste reduction programme as some may be too lengthy to be cost effectively pursued. ☐ Technique should be applied throughout a product's life cycle from raw material extraction to its production, use and final disposal to be effective.
Recycling – segregation at source	☐ Promotes staff involvement with the environment, and awareness and commitment levels maintained. ☐ Focus on the easier to separate items where a material value can be gained including paper, cardboard, aluminium cans and plastic drink cups. ☐ Combined with traditional methods of disposal, i.e. landfill or incineration, it is recognised that a well performing office can achieve between 60% and 70% recycling. ☐ Reduces the pressure placed on traditional disposal routes and can ultimately lead to cost reductions through reduced landfill tax costs.	☐ Sites may not have enough internal or external space for bin storage. ☐ Requires a strong commitment from all employees to ensure that recycling and awareness levels are maintained and contamination levels reduced. ☐ There may be cost implications for the use/hire of receptacles and collection of materials.

Continued

Table 4.4 *Continued*

Management option	Possible advantages	Possible disadvantages
Material reclamation facilities (MRF)	❑ No on-site segregation of different recyclables required. ❑ All materials sent to MRF therefore capturing recyclable that might have otherwise been sent for landfill/incineration. ❑ Separation of additional recyclables that are not currently segregated via any on-site segregation schemes, e.g. metals etc. ❑ Estimations vary on the amount of waste that can be captured via a MRF, but usually in the region of between 70% and 90% of general office waste can be recycled through this route. ❑ No requirements for multiple bins on site therefore minimising the space impact. Both segregated recyclables and general waste can be placed in the same waste receptacle. ❑ Contamination of recyclables collected is not important as with 'normal' office recycling segregation, as further separation occurs off-site at the MRF. ❑ Collection and disposal prices are comparable with other less environmentally sustainable disposal methods, e.g. incineration	❑ Accusation of encouraging 'waste apathy' among the public and the waste management industry. ❑ General requirement that waste is not compacted therefore increasing space requirements and vehicle collections – this can contaminate the waste and reduces the quality of recyclables removed. ❑ Remaining 30–10% of the waste, i.e. non-recyclables, are sent for landfill disposal. ❑ Costs are generally higher than traditional disposal options, e.g. landfill. ❑ Potential risks to operators in 'dirty' MRFs. ❑ There may be additional costs associated with increased frequency of collections.

Table 4.5   Template to calculate volume of waste arisings

Type of bin (a)	Typical weight (kg) (b)	Number of bins (c)	Number of lifts per month (d)	Monthly weight (b x c x d) = (e)	Annual weight (e x 12/1000)

not specifically weigh the bins or bags as they are loaded since the pick-up is usually one of several being performed. This can cause difficulty in obtaining direct volume rates, so a mechanism to estimate them should be used (Table 4.5). Some of the newer vehicles now provide weighing as an additional service, though this is charged for at a high premium.

For each bin used on site record its type (a); the waste company can provide, the typical weight of waste that the particular bin contains and the typical volume the bin is filled to when it is picked up (b). Record the number of bins on site (c) and the frequency with which they are picked up (d). Multiplying these figures will provide an estimation of both the monthly and annual weights. In this way an 1100 litre bin may hold 150 kg when full. If the site has four such bins picked up three times per week every week, the total waste generated is in the region of 90 tonnes per year. For each type of bin located on site, the expected weight, number of bins and number of lifts should be calculated.

On a regular basis, the typical weights of the bins (b) should be reviewed to ensure they are accurate, and the number of bins or lifts per month altered as and when they change. In this way the table becomes a working file which is regularly maintained as opposed to one which is visited irregularly.

Performance measurement The generation of raw data capturing volumes and costs through efficient monitoring can be used to identify trends in waste disposal and provide a proactive service to clients or customers. The information can be used to see annual or seasonal trends in volumes arising, special activities and alter the number of lifts required to remove the waste. This can have a significant reduction in the cost, since the majority of the costs incurred relate to the actual lift of the bin as opposed to the hiring of the bin or its disposal. Therefore the cost difference to remove a full bin or a half full bin is negligible.

The information can also be used to confirm continual improvement either as part of an environmental management system or through service level agreements. Traditional measurements outside of cost include:

❑ Total waste generated per occupant (or full-time equivalent);
❑ Total waste sent to landfill as a percentage of the total generated;
❑ Increase in recycling against previous year; and
❑ Total paper recycled.

Typical figures for waste generation in an office environment with a restaurant are 200 kg per person per year. Where a restaurant is not available, this value should be less. Again, within an office environment, 70% of waste generated can be recycled for a cost comparable with those of sending the waste to landfill, with 100% of paper being recyclable. For greater costs, more waste can be recycled.

In industrial and manufacturing environments, there is revenue that can be obtained from specific waste streams, specifically metals that can subsidise further recycling initiatives. Where high volumes of wastes are generated, such as plastics, it is recommended these are baled or compacted for sale, so generating an income.

Reporting There will be two elements to the reporting requirements, the first based on those requested by the waste contractor providing raw details, and the second using this raw data to provide performance measurements and trends.

Case study: Swindon and Marlborough NHS Trust – Carillion

The Swindon Great Western hospital has received much attention in the trade magazines, and awards in the delivery of sustainable principles in the design and construction of the premises. The private finance initiative (PFI) project was built by Carillion for £100 million and operated for 27 years. The hospital opened in November 2002 and showcased a number of innovative measures:

❑ 50% reduction of waste leaving site by designing out waste, recycling, prefabrication and composting biodegradables;
❑ 20 000 lorry movements reduced through improved earthworks;
❑ Sustainability adopted as a core work objective by all individuals; and
❑ Engagement with suppliers in a two-way process.

Following on from the innovative work undertaken, it seemed only natural and right that, as the service provider for the life of the contract, Carillion continued to build upon the sustainable foundations already laid. Since the opening, Carillion have been working in partnership with the Swindon and Marlborough NHS Trust to adopt a sustainable approach towards working at the Great Western.

This has involved the review and reduction of the production of waste, the use of energy and water; recycling wherever possible – cardboard, paper, aluminium, batteries and confidential waste. Future ventures will look at recycling wood, plastics, glass and, may be several years hence, even clinical waste.

The 2005 figures show that a total of 15% of the waste produced on site was recycled, a figure Carillion aims to increase by at least 5% in 2006. By working closely with their waste contractors and with improved segregation methods this is a very achievable target.

Energy and water saving initiatives for 2006 will start by raising awareness of the process by which the hospital is heated, cooled or otherwise provided by power and water and there will then be a series of roadshows telling people how they can help. Story-boards have been produced by the energy manager putting consumption into easy to understand terms – we use enough power per year in heating the hospital to heat and light Cricklade, a small town of 1500 homes, and, if we saved just 10% of that energy, the hospital could perform 38 more cornea operations or eight more coronary bypasses.

Working within the local and wider community is something that not only benefits those communities but also the individuals who participate in the activities:

❑ Fund raising within the catering department to enable Carillion to sponsor a child, Agnes in Kenya, leading to an improved standard of living, not to say better chances of survival;
❑ Sponsoring a local junior rugby team, promoting health and well being along with teamwork, collaboration and social interaction; and
❑ Forging closer links with the charity, Business In The Community, (BITC) to explore new ways of helping people, especially those who may be disadvantaged.

The information provided by waste contractors should be detailed as part of the contractual requirements, or requested directly mid-contract from the waste company. This is particularly appropriate where a contractor manages several waste streams either directly or through a series of waste contractors. Examples of information that can be requested include:

❑ Provision of changed or new documentation for carriers' licences or changes to disposal sites where waste is taken to;
❑ Cost of service over time including contracted and ad hoc service costs;
❑ Weight of waste disposed of segregated into waste type and location if applicable;
❑ Any incidents or issues arising over the service; and
❑ Initiatives or innovations to improve service delivery or increase recycling.

The following bullets points provide normalised environmental impacts of waste which can be used to translate typically dry waste figures into more meaningful data for staff to relate to:[8]

❑ One tonne of paper from recycled pulp saves 17 trees, 3 cubic yards of landfill space, 7000 gallons of water, 4200 kWh (enough to heat a home for half a year), 390 gallons of oil, and prevents 60 pounds of air pollutants;

---

[8] Waste facts and benchmarks – http://www.environment-agency.org.uk

❏ Each ton of newspaper recycled saves 4100 kWh of energy;
❏ Five out of every six glass bottles we use every year are thrown straight into the dustbin;
❏ Every tonne of glass recycled saves 1.2 tonnes of raw materials and the equivalent of 30 gallons of oil energy.

4.2.4   Waste management – auditing and performance management

The level of detail and effort from which a waste contractor is audited should be dependent on the level of risk the waste being generated has on both the environment and the organisation. In this sense, high volumes of hazardous or toxic materials or confidential materials pose a high risk.

(1) Where a high risk is posed, it is recommended performing a full review on an annual basis to provide satisfaction for the clear and compliant control of waste. The audit may also be performed when significant changes to the operations, activities, or collection/disposal sites occur;
(2) Where risk is low the following should be performed in conjunction with the waste contractor on an annual basis:

   (a) Existing and current transfer notes – either annual or for each individual transfer;
   (b) Existing and current carriers' licences as an original or clearly visible photocopy;
   (c) Checks performed by the waste contractor on receiving sites to ensure waste is adequately handled; and
   (d) Checks performed by the site manager to ensure that a waste contractor's carrier registration and/or site licence is still valid (they can be revoked by the Environment Agency before the expiry date). A phone call, fax or letter to the issuing department should be made and recorded.

Performing high-risk audits

The depth of the audit should be based upon a number of factors including the significance of the waste stream, previous audits performed and changes in legislation. Audits may be performed as part of an objective and target or through a tender bid.

Audits can therefore involve site visits, performing a paper trail assessment on the documentation to ensure the waste contractor is effectively managing waste to the waste producers' satisfaction and reviewing contractor site operations. A typical site audit should include:

❏ With a number of dockets/transfer notes, check the paper audit trail through from picking waste from a location through material acceptance, management on site and final disposal either on- or off-site. There should be a clear documentation transfer from one party to the next;

❑ Perform a site inspection looking for unbunded drums, open drains, etc. that could pose an environmental or health and safety hazard;

❑ Ensure all legal documentation is in place including discharge consents, site licences, etc.;

❑ If waste is passed on to a third party, review the complaint procedure for local community and check recent regulatory visit reports for any on-going issues or problems. Also review the process to prevent re-occurrences of problems which have arisen in the past.;

❑ Review the audit process used by the contractor to assess sub-contractor management and performance;

❑ Check the provision of training to ensure all contractor staff are competent; and

❑ Check the accidents/incidents process for contractor staff.

Once the audit has been completed, the auditor should provide an audit report describing the findings and key deficiencies to demonstrate the contractor's performance. The audit report is provided to the contractor for a response, and to provide feedback. This report will form the basis for the next audit to be performed and the regular meetings.

The recommendations of this audit will be sent to the contractor for comment, and to assign the appropriate corrective action. Dependent on the recommendations, a number of options will be open. The choice of action depends on the importance of the issue:

❑ Do not use contractor;

❑ Create action plan to work with contractor to improve rectifiable issues;

❑ Improve correspondence with contractor (relations);

❑ Offer incentives for improvement;

❑ Continue with periodic auditing;

❑ Award with more trade, financial rewards;

❑ Acknowledge efforts with 'certificate'.

In addition to an audit of the wastes leaving site, it is also important for a regular review of waste on site as part of the regular housekeeping checks to confirm the storage and efficiency of the current waste programme.

An internal review of waste storage, knowledge and transfer should be performed on a regular basis to ensure compliance is maintained and achieved. A typical proforma is described in Table 4.6 which can be performed as part of the general housekeeping activities. It is recommended performing this review a day prior to the uplift of bins to determine how full they are and therefore how well utilised. The results should be used to refine the waste service provided, by changing the equipment used and frequency of collections.

Table 4.6   Internal site reviews

Internal Site Review
Time: **Date:**   **Location:**
Inspection team members:
Signed off:

ACTIVITY	COMMENTS
Look inside some bins and skips	
– are the correct materials stored in the bins?	
– is there any hazardous materials?	
Is there any waste found outside the bins/ skips?	
Are all bins and skips either covered or in a drip tray to catch rainfall or spillage?	
Are all chemical waste bins locked and secure?	
Are all bins full or nearly full when collected by the waste contractor?	
Are staff aware of the waste disposal requirements?	
Are there any facts / figures of performance available?	
Are there any improvements that can be made to the waste management system?	

General Comments

Table 4.7   Waste management service level agreement

Waste management services

Prerequisites

(1) The service provider complies with all applicable statutory requirements and current codes of practice. All necessary certification for sub-contractors, sites and equipment is current and approved.

(2) The applied regime offers 'best in class' waste management services that minimise risk to the organisation business through meeting customer requirements.

Service level

(1) Service requests will be undertaken within agreed timescales with no disruption to the tenants and other occupants of the building.

(2) Additional requests and missed collections are being dealt with within agreed timescales and in relation to business criticality.

(3) Awareness programmes and roadshows will be developed and provided against agreed programme and timescales.

(4) Accurate monthly reports detailing:

  (a) Service delivery reports including describing missed collections;

  (b) Waste to landfill and material recovery figures;

  (c) Progress made on a site by site basis, and changes to sites as applicable;

  (d) Financial information indicating increases or decreases in expenditure.

(5) Waste disposal achieves agreed recycling and recovery targets.

(6) Tenant meetings, proposals and service reviews will be provided for the organisation and applicable tenants on an as-agreed basis to due timescales.

Service level

Table 4.7 details the service level the service provider is to adhere to when delivering the service within the contract area. The agreement is based upon a pre-requisite which must be applied at all times, and a series of service levels based upon performance from which the exact requirements necessary should be set. Therefore thresholds to determine what constitutes a late service requirement or missed request should be defined at the outset, along with recycling and recovery targets to be met over a given time period.

The service levels should be performed in the same way as any other – regular meetings and assessments to review progress.

Waste management, like any other service activity, requires investment to provide the optimum service performance, whether in new equipment for storage and bailing or through resources to provide the education and culture change. The provision of these resources will only be provided if costed as part of the service directly or indirectly. From personal experience, it has been beneficial to provide shared savings schemes, or the inclusion of additional fees for education to deliver the improved service provision. Paybacks on this investment are rarely over two years and commonly much less.

4.2.5 Resource optimisation – procurement and use

Currently, much of commercial waste globally goes to landfill. Disposing of potentially useful materials as waste means that the resources they contain are lost forever, producing a continuing demand on the need for non-renewable, raw materials. The focus for much of waste management has been the disposal of waste already generated – its collection, removal and recycling. The role of waste management however begins with the procurement of materials and their use to minimise the volume and type of waste generated (Figure 4.4).

At the procurement stage, the purchasing of recycled or recycled content materials help to provide a closed loop supply for the recycled materials. Encouraging the use of alternatives to packaging, particularly cardboard boxes and associated polystyrene will also significantly reduce costs and material usage. A typical example is the packaging around IT equipment designed to protect the items during transit. However, much of this is unnecessary – alternative packaging including plastic re-useable crates or blankets can be used. IT providers also provide equipment on pallets with shrink wrap for protection. Unfortunately, many IT providers will not offer this solution unless requested to.

When purchasing goods it is important to follow procurement guidelines and specify those that are energy- and water-efficient, non-polluting (or less polluting), durable, re-usable, recyclable, made from recycled materials and not over-packaged. These environmentally preferred goods do not have to be more expensive. Items with the lowest price may, in fact, cost more in the long run. It is important to understand that 'best value' is not always the lowest price, but includes a combination of whole-life costs and quality to meet user's requirements.

Areas to consider include running costs, indirect costs (such as the paper wasted by inefficient printers and photocopiers), administrative costs (such as the health and safety requirements for more hazardous substances), likely changes in legislation and disposal costs. Taking them all into account is the best way of achieving value for money (Table 4.8).

The main concept to improve resource efficiency is to ensure a greater level of centralisation and control in the services being provided. Initially, the focus should be on areas that can be directly controlled and managed to bring about change. This will help the ability to segregate at source and avoid the levels of contamination previously seen.

(1) *Reprographics* – Personal experience suggests approximately 40% of printed documents are either thrown away within five minutes or not collected, and therefore a reduction in this volume will reduce waste generated, but also the initial costs in paper purchase, energy costs, toner usage and wear and tear on the reprographics equipment. Consideration should be given to the procurement of multi-functional devices (MFDs) to enable printing to mailbox options.

Table 4.8 Resource optimisation through simple changes in procurement and use

Item	Goal	Specification examples
Stationery	To conserve resources, minimise waste and reduce pollution by specifying environmentally preferable stationery.	❑ Remanufactured toner cartridges from reputable suppliers who can guarantee quality and reliability. ❑ The density adjustment dial in printers be set to a high number, i.e. for a lighter print – it makes the cartridge last longer. ❑ Against spray adhesives, as much of the product can be lost to the atmosphere. Where necessary, go for water based or low solvent based products. Glue sticks are environmentally preferable, as there is less waste and they are non-toxic.
Paper	To reduce the amount of paper bought by using it more efficiently and economically. To buy 100% recycled paper, comprising at least 80% post-consumer waste. To increase the quantity of paper recycled.	❑ Photocopiers, printers and fax machines must: (a) cope with all the types of recycled paper used and can duplex print, i.e. produce double sided copies, without risk of jamming, and (b) have facilities for saving paper, e.g. controls on fax machines for stopping the print-out of transmission reports. ❑ Higher quality papers are only used when it is absolutely necessary and that standard, lighter weights of standard grade paper are sufficient for drafts, if possible, printed on both sides with a smaller type / font. ❑ Standard paper sizes are used in printing and design to prevent waste. ❑ Low-solvent, or solvent-free products such as water based glues and vegetable based inks and washes are used in keeping with the government's commitment to cut UK emissions of volatile organic compounds (VOCs). ❑ All reports commissioned are submitted on recycled paper comprising at least 80% post-consumer waste and preferably printed on both sides.

Continued

Table 4.8 *Continued*

Item	Goal	Specification examples
IT equipment	To select environmentally preferable products that have a lesser or reduced effect on human health and the environment when compared with competing products that serve the same purpose. Select products by taking into account the raw materials used in manufacture, durability, adaptability, distribution, re-use, operation, maintenance, packaging and disposal.	*Packaging* ❑ Apply the three Rs – Reduce, Re-use and Recycle. ❑ Specify that packaging is adequate to protect the goods and materials supplied, whilst avoiding over-packaging. Crates, pallets and, if feasible, boxes and cartons should be re-used. Cushioning and other forms of packaging which are not re-usable should be recycled. Put the onus on suppliers to remove their packaging: preferably for re-use; less preferably for recycling. *Disposal* (1) Reduce ❑ Check whether the purchase is necessary? Is there redundant equipment elsewhere in the department that can be used? ❑ Evaluate your requirements systematically and prioritise which items should be used. Ask the supplier for documentation that supports environmental claims. Keep abreast of latest developments. Look for goods that are durable and which can be easily upgraded. This is necessary to encourage manufacturers to end the wasteful practice of producing IT products that quickly become obsolete and so require replacing. (2) Re-use ❑ Check whether the IT equipment can be used elsewhere in the organisation, e.g. can it be used for less demanding work? Would it be possible to use the PC as a server?

		❏ Identify, evaluate and revise any standards or specifications unrelated to performance that provide barriers to the purchase of IT equipment made from remanufactured parts and recycled materials – see also section below on 'recycling'. Ask the supplier whether the manufacturer has a policy to use refurbished parts and recycled materials and the scale on which they are used.
		(3) Recycle
		❏ Request the supplier to provide details of the arrangements for the take-back of PC products and other IT equipment.
Timber	To use wood efficiently and encourage sustainable forestry practices which maintain the biodiversity, productivity and ecological habitats of woodlands.	*In maintenance contracts*
		❏ Waste timber and timber from temporary works should be re-used or recycled.
		❏ The number of applications of primer, under- and top-coats of paint or varnish to be used is minimised.
		❏ If painting or varnishing is required, preference is given to low solvent or water based media where technically suitable.
		❏ Maintenance and inspection routines should be supervised effectively.
		In supplies contracts
		❏ Where solid wood veneers are to be used, they are not less than 0.9 mm in thickness where commensurate with intended use.
		❏ Durability in the selection of fittings and furniture which should come with at least a five year warranty.
		❏ Inclusion with the product of full instructions for the care, repair and replacement of worn parts, including inventory numbers for parts and effective procedures for ordering them.

Continued

Table 4.8 *Continued*

Item	Goal	Specification examples
		☐ A preference for refurbished products or ones made from recycled material.
		☐ Plywood and particle board with the lowest formaldehyde level commensurate with the intended use.
		In cleaning contracts
		☐ No abrasive chemicals should be used on timber in furniture, fittings or joinery.
Hazardous substances	To minimise the use of substances which are hazardous to human health/environment and to replace them with substances that are known to be less hazardous or alternative techniques.	☐ Paints, glues, varnishes, preservatives, cleaning solutions and other materials supplied to, or used on the premises, conform to relevant UK legislation regulation on hazardous substances.
		☐ Suppliers avoid hazardous substances where safer alternatives exist.
		☐ Contractors carry out risk assessments under control of substances hazardous to health regulations (COSHH) and take account of instructions on product labels and refer to relevant guidance, for example, COSHH essentials, before they start work.
		☐ Regular replacement of ozone filters on printers and photocopiers according to manufacturer's instructions.
		☐ Plywood and particle board (and furniture containing them) should be manufactured only with adhesive binders containing a low level of formaldehyde, as high concentrations of this substance in ambient air can contribute to sick building syndrome. The binders should also be free of other hazardous substances.
Deliveries	To reduce the level of packaging waste arriving on site through over-packaged goods.	☐ Reduce packaging around catering and cleaning supplies which may not need to be so heavily packaged.
		☐ Review packaging for manufacturing products which arrive with substantial waste.
		☐ Identify packaging which can be re-used and effectively stored for this purpose.

A global pharmaceutical organisation moved their office base to a new location, which provided the opportunity to encourage a culture change in the environmental performance. The organisation historically had a number of desk-top printers used by most departments.

Multi-functional devices were provided at the new location for up to 70 people with a mailbox option to print to – a drastic change to the previous culture. All documents were printed to mailbox, the name and document selected to print the document. The ability not to print erroneous or missed documented reduced printing by over 40%, and enabled staff to feel they were contributing to the improved environmental performance on site.

It is anticipated there will need to be some awareness raised with staff to engage them into the process, with the environmental benefits promoted heavily, to reduce the volume of printing taking place.

(2) *Office areas* – Possibly the biggest area of change will be within the office areas where the removal of under desk bins, where a large quantity of recycled material is disposed, should be evaluated. It is anticipated that, in conjunction with the change in the reprographics set-up, the volume of waste generated can be greatly reduced, by up to 50% in some cases.

The replacement will be a communal bin comprising of paper, general waste and aluminium cans, with a separate bin for plastic cups. It is anticipated this will segregate and collect the main recyclables from site. The communal bins should be located in close proximity to each other to avoid their use as a general waste bin.

The removal of the under desk bins will also have an impact on the cleaners/janitors regarding the removal of waste. Certainly there will be a reduction in the number of bin liners required, and replacement bins as necessary. The removal of the bins – whether at night by the cleaners, or during the day by the janitors – needs to be resolved.

The following are key points to consider when looking at a centralised waste strategy:

❑ Ensure that floor waste stations blend into floor design;
❑ A strong behavioural/cultural change will be required to ensure the strategy works;
❑ Education of removal of under desk bins is crucial to strategy and the reasons for their removal;
❑ Leadership must start from the top – NO under desk bins to be provided;
❑ Waste bin types located in the waste hubs must be according to the main waste types generated;
❑ Main packaging waste from deliveries removed by service staff on or before delivery;
❑ Maximum recycling potential to be realised;

❑ Simplicity of service is key – if it is easy there will be greater buy-in; and
❑ Cleanliness and smartness of appearance of the bins.

(3) *Cleaning* – The cleaners collect much of the cardboard and the waste generated throughout the building and it is therefore critical in the implementation and maintenance of any waste minimisation programme. The cardboard must be flat packed when disposed of to the waste bins. This reflects good practice and it is recommended that the cardboard be removed to a cage rather than a bin to encourage the flat packing and recycling. Cleaners also have the ability to influence the lighting within the building, commonly as the last in areas within the facility, they can turn the lighting off.

(4) *Catering* – Historically there has been little in the way of recycling from catering other than cooking oil and aluminium can recycling. Deliveries to catering arrive with a vast amount of packaging that should be segregated for recycling. The removal of the cardboard from the catering areas to the waste areas should be performed. FM should look at the opportunity to compost the organic food to reduce waste disposed of.

(5) *Projects / FM* – The level of waste generated by projects is commonly not reviewed. It is anticipated there will be limited disposal of furniture, with re-use of any, either on-site or through charity networks, being the optimum.

(6) *Mail room* – There is a large volume of packaging waste that is deposited including cardboard and wood which can easily be removed and recycled.

(7) *Stationery* – A further area of improvement within the stationery activity is the level of cardboard packaging that arrives with the various items purchased. It is recommended these be maintained in a centralised point within the office areas for collection by the janitors on their daily rounds. FM should also look to purchase recycled stationery products to close the loop and encourage greater volumes of recycled materials.

4.2.6 Pollution control and management

With the introduction of liability for property owners and occupiers for contaminated land and affected biodiversity, the mitigation of pollution incidents and management of any emergency scenarios is an important issue. Aside from the obvious cost impacts associated with remediating contaminated areas, there is the potential legal action and associated publicity.

The potential liabilities to those managing buildings are:

❑ *Owner* – Liable for past poor practices with regard to site operations of others such as by the previous owner or tenant who cannot now be traced. Also liable for contamination which has occurred since ownership of their locations.

❑ *Tenant* – For historical pollution, if the original polluter cannot be found, the liability will rest with the landlord or owner of the building. However, tenants should ensure that their activities do not pose a risk to the environment to avoid the threat of future liability.

The following steps are recommended to ensure proactivity with the legislation and to identify any potential liability as a landowner, and reduce any liability as a tenant.

❏ *Owner* – A desktop condition audit of owned locations will identify potential historical contamination. This can be performed in a two-phase programme: initially a brief traffic light indicator study to identify high, medium and low risk areas through historical usage of land; and secondly a detailed desktop study to identify possible pollutants and pathways to environmental receptors for high-risk areas.

❏ *Tenant* – Review environmental compliance, both current and historically, to ensure no pollution potential exists. This will reduce the likelihood of being considered liable for contamination.

The following section provides some good practice measures to put in place to avoid pollution incidents in the first instance, and to plan for emergency scenarios if the worst happens. Best practice requires FM to ensure actions are taken to prevent, or mitigate the effects of polluting matter, or any solid waste entering drains, sewers or controlled (surface and ground) waters. This will include holding up-to-date consent to discharge trade effluent into the public sewer and its monitoring where applicable.

To reduce the potential for liability, there are a number of steps which can be taken:

❏ Maintain good relationships with the regulatory bodies and keep them informed of any planned changes;

❏ Fully document meetings, issues and changes to activities which are, or may become, controlled by regulation change;

❏ More importantly, ensure that contracts with service providers specifically include defined roles and responsibilities particularly for those involving use of hazardous materials such as cleaning and M&E services; and

❏ In the longer term, implement an umbrella management system that can be used to track and document all changes, thereby reducing the potential for liability.

Site drainage

No polluting matter is to be permitted to enter any drains, sewers or controlled (surface or ground) water. A good way to achieve this is to colour-code all drains and gullies, foul water drains (manhole covers and drainage grids) identified as red and all storm water drains identified as blue. All site personnel, including employees, contractors and sub-contractors can be made aware of the colour-coding scheme and the difference between foul and storm water drains. *Foul water*

*drains are designed only to carry waste water (trade effluent) to the public sewer system.
Storm water drains are designed to carry rainwater to local streams and rivers and other
controlled water.*

Comprehensive site drainage plans should be developed to identify the routes of
storm and foul water systems on site, the manholes and drains, and the location of
any separators, stop valves or the nearest stream. This is particularly important if an
incident or fire was to occur providing knowledge of coping with these incidents.
The plans should be reviewed and maintained on a regular basis.

Handling and storage

The transport of materials poses a potential risk and care should be taken when
storing and moving hazardous materials specifically. All materials deemed as
potentially harmful to the environment should be identified and stored in containers, tanks and pipes provided for this specific purpose. Where practicable dedicated
storage containers should be sited in bunded areas, covered, protected and secured
from tampering. Where dedicated storage facilities are unavailable, drummed substances (full or empty) are to be intermediately stored upon bunded palettes. There
is a legal requirement that all drums with a capacity of over 200 litres must be
bunded – this will include 45 gallon drums.

Regular housekeeping inspections should be undertaken to ensure that any
unidentified chemical drums are placed in a bunded area and marked for specialist
attention. In addition, it is recommended that regular inspections and maintenance
checks are undertaken to ensure the integrity of the dedicated storage facilities and
bunded areas.

Diesel deliveries / chemicals

Deliveries of diesel or in fact any bulk chemical poses a significant potential risk,
both from an environmental perspective and a business risk if fuel is not available
within the short term. It is essential that the risk of a major spill is assessed and
incorporated into any business continuity plan.

When contracting out the service, ask the company if all drivers have received
training of what to do in the case of a spillage, and whether spill kits are carried on
board the lorry. Whilst this seems logical, it is surprising how many do not have
kits, and have never received training. When a major delivery is being made, ensure
a member of the FM team is available to support and watch the delivery to avoid
spillages whilst coupling or de-coupling, and to know where the spill kits are, if
required. Provide drip trays if necessary to be kept under the tanker coupling to
catch any drips.

If a major incident does occur, resulting in chemicals entering into the drainage
system, contact the relevant authority and make them aware of the situation. It
is an offence if these bodies are not contacted within a short time of the incident.
Look to shut off the drainage system to contain the spillage and limit any chemical
escaping into the environment. The use of up-to-date drainage drawings will help

to identify the route of any spillage and where any containment within the existing drainage system can be held.

Spill kits

The provision of emergency spill kits should be regularly maintained in preparation for potential spillage incidents. It is also essential that personnel have adequate knowledge and competence to capably deal with spillages both within and outside normal working hours. A documented report mechanism should also be set in place to detail the results of the incident and investigate root causes to ensure preventative measures are taken to avoid another incident in the future.

Site survey

A site survey should be conducted on a regular basis to identify all locations where there is the potential for spillage incidents to occur. Where spill kits are already in place the survey should assess the appropriateness of the kit types and materials allocated to the designated locations. As a result of this survey, a location map should be developed identifying the various kits across the location, and held in key areas such as the maintenance office or the security office.

Training and awareness

All personnel liable to be involved in a spillage incident are to be made aware of emergency spill kits, how to use them and what to do in an emergency. All staff should be made aware of whom to contact given an emergency. Guidance notes should be posted with each spill kit for reference, which denote the order in which actions should be taken to ensure personal safety and prevent the occurrence of an environmental impact. Emergency contact numbers should also be provided.

4.2.7   Water management

Water management is commonly part and parcel of utilities management with a review of consumption and savings tied into that of gas and electricity consumption. Whilst the management of water is generally performed through an engineering department, the use of water and how it is controlled is very different to that of gas and electricity with different factors which need to be understood.

FM commonly manages all aspects of water usage and disposal from reading the incoming meters and any sub-metering within the location, managing the infrastructure of water supply to toilets, restaurants, fountains, vending machines and the disposal of waste waters through either sewerage or storm water drains. Industry uses water in varying degrees – many clean industries such as printed circuit boards manufacturing are using cleaner solvents and therefore the water disposed of is well below any legal limits. However, since this is classified as an industrial process it still requires legislation and the confirmation through a discharge consent. Heavier industry using large quantities of water have invested in settlement tanks or filtration equipment to reduce the costs of discharge and risk of non-compliance.

This section describes briefly the various steps to understand water consumption within the portfolio, and set in place the resources and measures to achieve a reduction in water usage and consumption:

(1) *Determine the costs of water services* – Look back at the invoices paid, which will include the standing charges, fees payable based on volume used and passed to sewers. If a discharge consent is required, charges may be higher for disposal costs. Identify the amount of time spent on maintenance, the cost of consumables such as water softeners, and any further costs to manage the system. This should provide a total annual cost for the waste service comprising direct and indirect costs.

(2) *Appoint water champion and project team* – Choose a team of individuals who are given the responsibility and resources to manage the project to identify and implement water saving measures. Ensure there is a senior level sponsor for the project with clear and defined deliverables set for the team, reporting timelines and structure.

(3) *Measure usage* – Set up a programme to measure existing consumption and to confirm water bills provided. Where accurate, water bills can be used for historic consumption. Profile the results in tabular form to make assessing consumption and reviewing trends easier. This should be performed for water into site and the volume of water discharged. Where possible meter readings should be taken on a monthly basis where consumption is low, or weekly if the usage is much higher. This will provide a greater level of clarity to identify trends and see excessive usage from leaks.

Typical areas of high consumption within offices and retail units are from toilets together with associated urinals and wash hand basins (Figure 4.5). What is noticeable is the relatively low water consumption from a restaurant facility in comparison.

(4) *Identify savings* – The next stage is to identify where water is being consumed within the building and to reconcile that with the anticipated or predicted volumes based on meter size, sub-metering or typical industry figures. A graph can be generated to represent the findings (Figure 4.5), which highlights clearly where the greatest focus should be made. In this example, excess usage can be seen in all areas, with a particularly high consumption in toilets and the restaurant. By concentrating on these areas, the largest savings can be made for the smallest effort.

Much of the data required will need to be estimated based on knowledge of the units within the building. An example is shown in Table 4.9 to calculate the water consumption from WCs, urinals, taps, and showers – a similar methodology should be used for all water appliances including those in the restaurant. For each individual appliance, identify how many there are (*a*), the water consumption of each (*b*), and roughly the amount of time each appliance is in use for (*c*). For toilets and urinals, this will be based on an assumption of the flush rates and number of people within the building. By multiplying these figures together a rough estimate on water consumption can

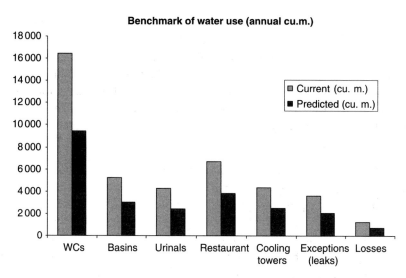

Figure 4.5 Typical consumption versus predicted consumption.

Table 4.9 Template for the collection and calculation of water data. (Source: adapted from Environment Agency, *Waterwise*.)

Item	Number of units [a]	Flow rate (l/flush or l/min) [b]	Operating time (min/day or flushes/day) [c]	Water used (l/day) [a × b × c]	Comments
WCs	30	9 l/flush	5 flushes/day	1350	
Urinals	10	7.5 l/flush	7 flushes/day	525	Office hours only
Taps	50	4 l/min	3 min/day	600	
Showers	03	10 l/min	10 min/day	300	

be provided. A total of 80% or more of the water consumed should be identified and confirmed through reconciliation with the water bills and metering within the building.

Once the highest consuming activities and those with the greatest potential for reduction have been identified, there are a number of ways to reduce water consumption:

❏ Detecting leaks in supply pipes should be performed either by the local water company or a specialist contractor. Leaks can account for up to 20% of consumption and lead to structural problems if unidentified;

❏ The maximum flush volume on toilets can be reduced to 6 litres for newly installed toilets, although older models should be maintained at 7.5 litres due to differences in bowl dynamics;

❏ Waterless urinals offer the same service as traditional urinals when used and maintained correctly. It is recommended trialling these units for a period of time to the manufacturer's instructions – waterless urinals are in operation at airports and have posed no difficulty;

❑ A variety of mechanisms are available for the collection and storage of rainwater to be used on site, with a number of proprietary systems available to provide a complete system. The water can be used primarily for toilets; and

❑ The use of greywater is common in Japan and Germany where it has been in use for the past 20 years.

(5) *Set benchmarks* – Typical water consumption figures for an office environment are in the region of 15–20 m^3 per FTE (full time equivalent), inclusive of visitors, based upon previous experience. Best practice figures of below 8 m^3 per FTE have been set, with the UK Environment Agency having achieved 7.7 m^3 through the efficient use of greywater and education of staff. Benchmarks should be set based upon realistic targets and a project plan. A site using 20 m^3 per FTE could look to set a 10% year on year reduction target over a three year timeframe.

(6) *Develop a project plan* – Based on the various projects identified, develop a water management plan, classifying the no-cost, low-cost, and the medium-to-high cost options, with payback periods. This will provide a priority list for the implementation of the various projects and highlight where the greatest workload is going to be. In addition, the resources required and timescales should also be identified for each of the projects to ensure a burden is not placed on individuals, and to effectively plan the savings. A template is provided in Table 4.10 which describes the various elements to be identified, and a normalising factor of carbon dioxide (CO_2) from which to provide performance measurement data.

It is also essential that all calculations and raw data are maintained for the various savings, and the steps taken to provide the final savings. This will be required when verifying the figures, and to ensure that the knowledge generated at this stage is not lost in future. The lack of sufficient data is a common hindrance to be able to verify savings and can lead to doubling of efforts and lost time.

(7) *Implement water management plan* – Once a plan has been developed, it must be implemented according to the timetable and programme. The plan must be maintained and reviewed regularly, and where necessary, altered to maintain progress.

At this stage, communication is the key to engage staff and to maintain their interest, which can be achieved through a number of mechanisms including:

❑ Introducing a staff suggestion scheme to encourage new ideas and initiatives for saving water;

❑ Reporting of wastage from dripping taps and overflowing WC cisterns;

❑ Displaying posters and notices requesting water taps and appliances such as vending machines and tea boilers should be turned off when not required; and

Table 4.10 Template for the capture of water efficiency savings

Proposed project	Location	Est. cost (£)	Est. savings (£)	CO_2 (tonnes)	Payback period (years)	Priority	Org. actioned	Project manager	Date work ordered	Expected completion date
1										
2										
3										
4										
5										

❏ Progress reports based on savings achieved through the intranet, posters, newsletters or in-house journals.

(8) *Review performance* – On a regular basis, a review should be undertaken to verify the savings achieved, projects completed and the consumption figures. This will also help to identify the lessons learnt and make implementation of further savings easier. At this stage, a review should also be made of the meter size that may be changed to a smaller size due to the reduced flow that also carries with it a lower charge from the water authority.

It will also be useful to publish savings achieved and reduction in consumption rates dependent on the audience, based on the performance measures identified within number (4) above.

4.2.8 Air emissions

Traditionally air pollution has been viewed as an issue affecting heavy industry predominantly and of little concern to the vast majority of businesses. However, recent legislative changes, particularly relating to solvents and ozone depleting substances, together with increasing numbers of complaints from the local community mean that business needs to ensure compliance. Air pollution originates from four main sources:

(1) *Combustion processes* in the form of smoke or combustion gases, commonly discharged through a chimney from boilers;
(2) *Chemical releases* from emissions can be a by-product of heavy industry, though these will be regulated under tight legislation. More commonly, chemicals are accidentally released such as ozone depleting substances;
(3) *Evaporation* predominantly from the use of solvents in cleaning chemicals, paints and thinners. For many of these substances, there are safer and more environmentally friendly alternatives;
(4) *Particle discharge:* dust and particles can arise from both part of the industrial process or from project activities dealing with external building works. Dust can be not only a hazard to health, but also a major nuisance to local residents and neighbours.

This section will provide information on how FM can manage and mitigate air pollution. There are a number of mechanisms that facilities management can provide to ensure effective control over air emissions and to reduce the impact on the environment and local community.

Flue gases

Boilers should be maintained to their maximum efficiency through a planned maintenance programme and the use of a computerised system, where appropriate, to enable the boilers to burn at their most effective rate. This provides not only energy savings but minimises the exhaust of combustion gases emitted. A contingency plan should also be developed to provide back-up should the system fail.

Volatile organic compound emissions

Emissions of volatile organic compounds (VOCs) are generally very low and specifically concern the use of solvents in paints, thinners and cleaning chemicals. Alternatives to solvent-based products are available and should be used as substitutes. This will mean solvent free products do not necessarily have to be stored in flammable proof cupboards, and may not need to be disposed of through the hazardous waste route, saving both costs and handling charges.

Local exhaust ventilation

Facilities managers are commonly required to manage fume cupboards and other exhaust ventilation, predominantly in manufacturing and laboratory environments. These filtration units should be set up and monitored regularly through a planned maintenance schedule by a competent specialist contractor. Up-to-date and complete records, logs or service sheets should be kept of all inspections, monitoring and maintenance activities as part of the schedule. A report should be provided by the contractor, detailing the status of the units and specifically highlighting those containing defects on an annual basis.

The defects are categorised into three sections covering the usability of the units. Units should be deemed to be unusable where they provide a hazard not only to the user, but also where the removal of air is insufficient, or damage to the environment will occur if the filters are not removing the chemicals as required. In both cases the units should be stopped, and in the latter case where excessive air pollution may have occurred, the incident should be reported to the regulator.

4.2.9   Community engagement

What is being a good neighbour? All organisations are part of the local community in some way or another, and have an impact. Managed well, this impact can provide positive benefits to all parties. Business has a positive impact on the community because of the employment the company provides and may produce goods and services which can be consumed locally.

A company can add an extra dimension by taking the initiative and supporting the community through a programme of corporate community investment. This programme can take the form of charitable donations, staff volunteering and providing other resources, such as the use of professional skills or the use of equipment and premises. The opening of facilities meeting rooms and general space during weekends and evenings to support local groups can maximise the use of the facilities for a relatively small cost and improve local relations and the standing of the organisation.

Of course a business may also have a negative impact. This might include disruption and noise from the day-to-day operation of a site and poor relationships with key local stakeholders. A good community programme should go hand in hand with measures to keep these negative impacts to a minimum.

The role of community relationships varies widely from organisation to organisation, with human resources or public relations departments taking the lead in many larger businesses. For smaller organisations, it may become part of everyone's responsibility. There is certainly a role for FM to support the local community either directly for its own staff, or to co-ordinate a wider programme in keeping with the organisation's corporate policy.

It is necessary for any organisation to consider their position within the community and the investment that can be offered:

❏ Understand the current community initiatives taking place on site through the various departments;
❏ Determine the potential for community support that would maximise the investment made – time, capital, equipment or resource based, or a mixture of these elements;
❏ Determine the length of time and commitment to be provided, i.e. an on-going programme of works, or a one-off activity;
❏ Understand the legal requirements covering the disclosure of community investments, particularly linked with the tax breaks available;
❏ Determine the current and planned projects of local trusts and charitable bodies to link in with existing projects where support is required; and
❏ Identify how the investment will be monitored, measured and reported.

To be put into practice, each of these investments needs to be treated as any other business idea, with a business case, strong process and good management. The mechanisms to measure the success of the project will be critical not only for its immediate success, but also for future projects. Measures of success may be through awards or external verification. This will involve collaboration with the local community on an on-going basis to provide continual improvement and to ensure the programme is effective.

Community investment programmes[9]

American Express registered big positive increases in staff satisfaction from groups who had been volunteering in a local school, compared with a control group of non-volunteers.

NatWest's Board Bank scheme, which encouraged volunteers to provide arts organisations with a skilled manager for boards, showed good benefits for the business as well as the community groups. From the volunteers, 92% were shown to have picked up new skills, with 77% improving their overall confidence and similar improvements for problem-solving and meeting skills.

Marks and Spencer evaluated the skills gains achieved by staff from their community assignments. Impressive gains were shown across a range of competencies, including teamwork, decision-making and leadership, project management and customer focus. Overall, both managers and the individuals rated around a 30% gain across all the competencies.

---

[9] Business in the Community, Business Impact Project – Community November 2000 – http://www.business-impact.org

A series of examples are provided in the information box highlighting that the knowledge and skills organisations take for granted have a high value in the local community, and particularly for highly skilled activities such as engineering, project management and finance. This is where FM could play a significant role through the provision of its skilled labour as part of a community investment programme.

4.2.10   Biodiversity

Biological diversity is the term that represents the richness and variety of all living things that exist throughout the world, including species, habitats and ecosystems. Biodiversity conservation is similar to the more traditional idea of 'nature conservation', but it also implies that social, cultural and economic values are important considerations. Human society's interaction with biodiversity determines whether our economic and social development is sustainable.

The increasing numbers and size of facilities occupies space once inhabited by plants and animals and erodes the natural environment, replacing it with urban areas and man-made 'green' spaces. This is leading to much of the natural environment becoming protected by law including both areas of land and individual species. Many organisations are looking to put something back for the benefit of the biodiversity so that not only the species, habitats and ecosystems will gain but also the facility occupants and local community.

Many organisations are developing biodiversity action plans (BAPs) as a means to deliver biodiversity conservation programmes capturing the whole process, or how it will be achieved. The development and implementation of a BAP is similar in many ways to a management system (Section 2.4).

The development of a BAP should be performed in conjunction with local biodiversity groups who can offer support on local plants and species, and align with existing programmes taking place in the region. If an on-site programme is inappropriate due to the size and location of the facility, consider providing support to a local project. The development of the BAP should include clear objectives and targets including the means to measure progress and show continual improvement. The benefits of providing green roofs will increase staff and local community's knowledge of how biodiversity relates to business, and can attract nesting birds and unusual sitings.

Some measures to include either within a BAP, or to deliver as part of a grounds management programme, are provided below to maintain biodiversity, minimise impacts and to promote the benefits:

- ❑ Consider incorporating courtyards and/or roof gardens into the design;
- ❑ Landscape open spaces sympathetically to the needs of both wildlife and building users;
- ❑ Use local materials (stone, etc.) to maintain the local character;
- ❑ Consider the local flora and fauna, and plant native species to enhance local biodiversity;
- ❑ Use materials only from sustainable sources, such as peat-free compost;

❏ Maintain to ensure continuing support for local biodiversity;
❏ Compost or chip cutting and materials on site for re-use as natural fertiliser;
❏ Use bio-degradable chemicals with minimal toxicity to wildlife; and
❏ Use waste water or storm water run-off for watering plants and shrubs.

4.3   Occupant satisfaction

Much of the focus on developing and delivering a 'green building' is based upon the physical attributes of the facility and the direct asset value. A facility should also provide an environment to encourage and satisfy employees, visitors and customers within the facility. The combination of both factors is difficult, and can be in conflict with each other through issues such as ventilation rates.

There has been a significant level of research into the provision of adequate space, services and equipment to improve satisfaction and thereby increase performance. Whilst the improvement of the environment does have an impact on productivity, it is difficult to ascertain the true value outside of other changes and influences. However, the performance of an end-user satisfaction survey can identify ways to improve the working environment at no cost and to minimise calls to helpdesks.

The satisfaction of occupants will also include the FM staff and contractors working on site, providing suitable working hours and pay in keeping with human rights. Cutting costs on resource intensive contracts such as cleaning or catering will affect the pay and working conditions of staff.

Section learning guide

This section looks at the impacts on occupant satisfaction and mechanisms to identify and provide effective space for improved productivity. This includes identifying the points for changing space requirements and occupancy evaluation:

❏ Introduction to occupant satisfaction;
❏ Triggers for a change in the workplace;
❏ Pre- and post-occupancy evaluations; and
❏ Templates and proformas.

Key messages include:

❏ Procurement of goods and services is one of the greatest environmental and social impacts from FM; and
❏ Local contractors and suppliers should be considered.

4.3.1   Occupant performance

A workplace review is not simply a case of reorganising space or ensuring regulatory compliance. It represents an important strategic task which can fundamentally change business performance by maximising efficiencies, enhancing productivity, promoting positive behaviours and successfully communicating your values and brand. Almost three-quarters of property professionals feel that the workplace has a role to play in helping companies achieve their business plans, but only 63% are convinced that the board recognises the importance of the workplace and less than half are satisfied that enough is being done to make the workplace contribute positively to the achievement of business objectives.[10]

Poorly designed offices are having a major impact on productivity by as much as 19%. The design is also linked to job satisfaction, recruitment and retention, with four in five professionals considering the quality of their working environment very important to job satisfaction and more than one third stating that working environment has been a factor in accepting or rejecting a job offer. Over half of professionals believe their office has not been designed to support their company's business objectives or their own job function. Personal space (39%), climate control (24%) and daylight (21%) are the most important factors in a good working environment according to those surveyed.[11]

Good office design can increase productivity by nearly 20% – and is a crucial factor in job satisfaction, staff recruitment and retention. The cost of providing accommodation for office workers is dwarfed by the cost of their salaries – the impact of the office on staff in terms of increased productivity and effectiveness will have a much greater financial impact than the factors influencing the cost of office accommodation.

A report by PricewaterhouseCoopers[12] for the UK Education Department found a 5% improvement from high quality school buildings where sustainable improvements had been incorporated. Areas highlighted included open areas for play; light colours to create a sense of warmth; community involvement; daylight and natural ventilation; and a move away from narrow dark corridors.

In the United States, the 'West Bend' study by Walter Kroner[13] documented productivity gains from day lighting, access to windows, and a view of a pleasant outdoor landscape at the West Bend (Wis.) Mutual Insurance Company. The performance of clerical workers in a new building, opened in 1991, was compared to that of workers in an old building to see which group could produce more reports in an allotted time. Employees in the new building were also supplied with individual controls that allowed them to adjust temperature and other conditions in their work environments. According to the study, productivity gains in the new building increased by 16%, with the personal controls alone accounting for a 3% gain. The building also reduced energy consumption by 40%.

---

[10] Performance Measurement in Real Estate Usage, part of the Directors Briefing set, Haywards 2005 – http://www.haywardsltd.co.uk

[11] Gensler's These Four Walls: The Real British Office 2005 – http://www.gensler.com/

[12] *Building Performance: An Empirical Assessment of the Relationship Between Schools Capital Investment and Pupil Performance*; PricewaterhouseCoopers; ISBN 1 84185 402 6 – http://findoutmore.dfes.gov.uk/2006/08/school_building.html

[13] Kroner, W.A., Stark-Martin and Willemain, T. 1992. *Using Advanced Office Technology to Increase Productivity. Rensselaer Polytechnic University*, Center for Architectural Research.

Another report is the Heschong Mahone Group study 'Daylighting in schools',[14] which was conducted on behalf of the California Board for Energy Efficiency. The researchers analysed test scores for 21 000 students in 2000 classrooms in Seattle; Orange County, California; and Fort Collins, Colorado. In Orange County, students with the most daylighting in their classrooms progressed 20% faster on math tests and 26% faster on reading tests in one year than those with the least daylighting. For Seattle and Fort Collins, daylighting was found to improve test scores by 7–18%.

A study of windows and views in seven buildings in the Pacific Northwest[15] found that employees in work areas with windows were 25–30% more satisfied with lighting and with the indoor environment overall, compared to those with reduced access to windows. Window views may be especially effective in providing micro rest breaks of a few minutes or less, which have positive impacts on performance and attention.[16]

4.3.2 Changes in work patterns

The changes in work styles and patterns are leading to a move away from the traditional one person to one workstation desking arrangement. Staff are working part time, with shared roles, flexible hours, flexible locations and some are home based. The workstation share ratio provides an indication of desk sharing or alternative ways of working when employed at a particular location. A typical desk share ratio for flexible workspaces is 1.3 : 1, i.e. 1.3 occupants for each workstation, with the aim of providing:

❏ 1 : 1 for fixed static staff;
❏ 3 : 2 for flexible static staff;
❏ 3 : 1 for mobile staff; and
❏ 7 m^2 net useable space per office.

The provision of the working space should encourage the collaboration of staff, time spent away from the desk and computer and greater interaction with other staff. The various types of space should support this flexible approach. This will include an open-plan space encouraging managers to be located together with staff, and a suite of meeting spaces both formal and importantly informal:

❏ *Open-plan* – the team workspace is the fundamental work unit for the majority of staff. This space provides individually owned space with additional shared facilities to support daily work activities;

---

[14] Heschong Mahone Group study 'Daylighting in schools', – http://www.h-m-g.com/projects/daylighting/projects-PIER.htm

[15] Heerwagen, J., Loveland, J. and Diamond, R. 1991. *Post Occupancy Evaluation of Energy Edge Buildings*, Center for Planning and Design, University of Washington.

[16] Zijlstra, F.R.H., Roe, R.A., Leonora, A.B. and Krediet, I. 1999. *Temporal Factors in Mental Work: Effects of Interrupted Activities*. Journal of Occupational and Organisational Psychology, 72: 163–185.

❏ *Personal office* – offices should be discouraged where there are quiet rooms and small meeting rooms which could be used. Where necessary, an office should only be provided for staff with a managerial role whose job requires working on sensitive matters which require a level of confidentiality. Offices should be located towards the central core of the room so as not to restrict the natural daylight and be a shared space when not in use;

❏ *Meeting room* – a range of rooms, that is bookable and available to all staff. Larger meeting rooms can also have the facility to be used as training rooms if sufficient data and network points and equipment are available for the users;

❏ *Breakout space* – these spaces are provide for informal meetings and breakout intended to promote networking and encourage interdepartmental interaction. They also provide core services such as reprographics, library facilities and stationary around which social interaction is encouraged. A nominated person should be made responsible for these hubs to ensure they are maintained;

❏ *Quiet room* – non-bookable space for individuals to work in privacy enabling focus and concentration. Layout should allow for up to two quiet users; and

❏ *Restaurant* – this is the main communal facility providing a series of spaces for breakout and meals/snacks. This includes the provision of soft furniture promoting informal breakout/meeting space to take the pressure off formalised meeting spaces. Loose furniture allows the flexibility to create large meeting spaces for the whole organisation (and the provision of adequate connectivity allows audio-visual aids). Furniture should be provided that allows small group meetings and large group gatherings at meal time.

Triggers for a change in workplace

Costs associated with changing the workplace and the disruption caused mean that this option should only be taken when necessary. There are a number of trigger points to identify the appropriate time to provide the change, to review their space utilisation and move towards the best practice model.

Organisational changes

❏ Organisation redesign based upon changed focus or realignment to business activities;
❏ Recruitment/retrenchment of staff;
❏ Management change within location;
❏ Mergers and acquisitions – to promote integration and culture change; and
❏ More flexible global workforce – to increasing mobility.

Cost management initiatives and activities aimed at reducing cost:

❏ Minimise churn activity through the provision of efficient space first time and reducing cost of change; and

❏ Rationalisation of portfolio due to lease expiry, break clauses or mechanisms to alter space usage.

Changes in available technology, enabling a reduction in the number of desks, the occupant footprint or the desk footprint. This may include flat screen technology, use of laptops, etc.

Occupational health changes to the use of the workplace to provide an effective space for personnel to work.

4.3.3   FM staff satisfaction

An important aspect commonly missed during the review of performance on site is the FM staff themselves, and in particular the operational staff who may be sub-contracted. Commonly these staff will provide the cleaning, catering and security on site, and usually there is a relatively high turnover of these staff in comparison with the rest of the business.

Many of these staff work irregular hours to provide the income. FM has the opportunity to improve the employment opportunities and respect for these staff by integrating them into the team as a whole and treating them fairly. This may mean providing them with a higher wage to reduce turnover of staff, or simply starting the cleaning shift in the evening, e.g. 6 p.m. to reduce the need for night-time working. Importantly, this will not only integrate the teams, but also enable the site to be closed and lighting turned off overnight.

A pharmaceutical organisation moved its cleaning regime forward to commence at 6 p.m. enabling cleaners to feel part of the FM team. Through liaison with the senior managers who were commonly still working at this time, a personalised service was provided improving customer satisfaction, staff enjoyment and an overall improved service. A higher than local wage is provided which significantly reduces the turnover and saves on the need to retrain staff on a regular basis.

Within tenders and contracts it is important to measure the satisfaction of the FM contractor staff as well as the employees in the building – whether this is through a satisfaction survey or as part of the appraisal process will differ from company to company.

4.3.4   Pre- and post-occupancy evaluations (POE)

An occupancy evaluation involves the systematic evaluation of opinion about buildings in use, from the perspective of the people who use them. It assesses how well buildings match users' needs, and identifies ways to improve building design, performance and fitness for purpose. An occupancy evaluation is best performed prior to and following a project or move to determine the change in satisfaction. The pre-evaluation will also help to identify the aspects staff in general are uncomfortable with and those areas they enjoy, and incorporate these lessons into the revised project output. The success of these changes will be measured through the post-evaluation.

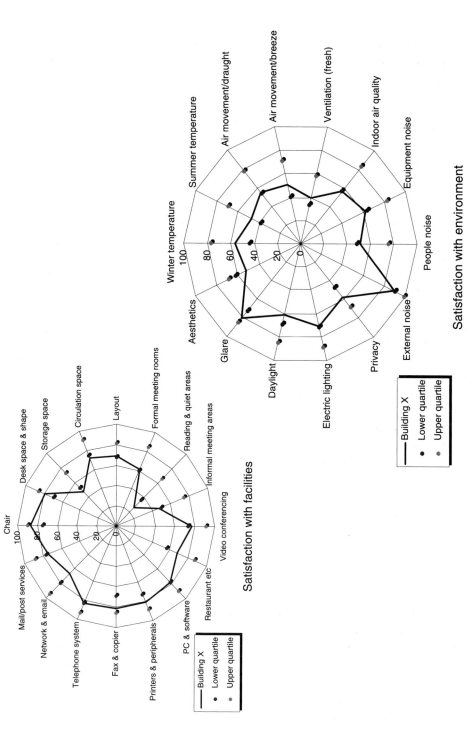

Figure 4.6 Typical outputs from a pre- and post-occupancy evaluation.

The POE can be performed across all sectors within the business, including FM staff, to gauge feeling and satisfaction with their own working environment.

The POE is performed using a questionnaire to gain a direct feedback from the occupants, and uses these experiences as the basis for evaluating how a building works for its intended use. It can be used for many purposes, including fine-tuning new buildings, developing new facilities and managing 'problem' buildings. Organisations also find it valuable when preparing for refurbishment, or selecting accommodation for purchase or rent. Most importantly, the tool allows occupants to provide direct feedback on the performance of the building and how it meets their needs.

A typical output of the POE is described in Figure 4.6 for two assessments which were performed – one on the facilities provided and the other on the environment. The 'facilities' covers questions related to satisfaction with items such as chairs, storage, desk, restaurant and printers. The 'environment' covers issues including temperature, ventilation, glare, daylight and noise. Occupants were asked to rank each of these various aspects from excellent through to very bad which enabled a consolidated score to be provided. Using similar locations a benchmark can be provided to compare performance with peer companies.

From this initial point, where the current satisfaction is known, the project can incorporate the lessons in terms of areas to change and those to improve. From the figure, the critical areas to change would include storage space and reading/quiet rooms from a facilities perspective, and air movement and lighting from the environmental side. The inclusion of occupants and listening to their requirements also improves performance simply because issues raised are resolved.

Incorporated within the POE could be a review covering focus groups – small teams who represent the occupants where a greater level of dialogue provides further feedback on the performance and what may or may not work well. A typical POE has three phases:

❏ *Preparation*: Identification of user groups, timetabling, selection of participants, letters of invitation;
❏ *Interviews*: Small groups of like users are interviewed while walking through the building, which provides the prompt for their comments and observations. A review session is held to verify comments, establish priorities and review the process. Observation studies and written questionnaires may also be used; and
❏ *Analysis and reporting*: Documentation of participant findings, generation of recommendations, compilation of a report and presentation. Comparison with the pre-evaluation to demonstrate improvements.

List of websites

The details below provide a full list of the various documentation and websites available for further information. This list is by no means exclusive, but will certainly provide an initial starting point for further information to support the various topics discussed.

Life cycle or building / products

Environmentally Preferable Products and Services (EPP) Scientific Certification Systems (www.scs1.com)
Forest Stewardship Council (www.fsc-uk.info)
Green Building Lifecycle Assessment (www.eiolca.net/index.html)
Green Label Testing Program, Carpet and Rug Institute (www.carpet-rug.com)
Green Seal (www.greenseal.org)

Energy management and climate charge

ActionEnergy provides a range of energy efficiency best practice resources (www.actionenergy.co.uk)
Building Research Energy Conservation Support Unit (www.bre.co.uk/brecsu)
Carbon Disclosure Project (www.cdproject.net/)
Carbon Trust (www.thecarbontrust.co.uk)
EEBPP (BRECSU) (www.eebpp.org.uk)
Emissions Reduction Programme by US States (www.ens-newswire.com)
Energy Efficiency benchmarking (www2.energyefficiency.org/default.asp)
Energyroadmaps.Org (www.energyroadmaps.org)
Energy Saving Trust provides guidance (www.practicalhelp.org.uk)
EnergyStar from US EPA (www.energystar.gov)
EU Emissions Trading Scheme (http://europa.eu.int/comm/environment/climat/emission.htm)
Kyoto Agreement, European Programme (http://europa.eu.int/comm/environment/climat/kyoto.htm)
Renewable Power Association (www.r-p-a.org.uk)
UN Framework Convention on Climate Change (http://unfccc.int/)
University of Southern California (www.usc.edu/uscnews/stories/11608.html)
US Emissions Programme (http://en.wikipedia.org/wiki/Kyoto_Protocol)

Use of resources

Environwise (www.envirowise.gov.uk)
UK Water (www.water.org.uk/)
Waste and Resources Action Programme (WRAP) has a range of activity on construction
  waste (www.wrap.org.uk)

Marketplace

Best practice benchmarking (www.dti.gov.uk/mbp/bpgt)
Chartered Institute of Purchasing Supply (www.cips.org)
Environmentally preferable purchasing from US EPA (www.epa.gov/opptintr/epp)

Biodiversity

Business and biodiversity (www.businessandbiodiversity.org/)

Facilities Management

British Institute of Facilities Management (BIFM) (www.bifm.org.uk)
Corporate Real Estate Network (www.corenetglobal.org)
Facilities managers directory/ Facilities managers association (www.fmd.co.uk)
Facility Management Association of Australia Ltd (www.fma.com.au)
Global FM (www.globalfm.org)
International Facility Management Association (www.ifma.org)
MBD Ltd, March 2006 (www.mbdltd.co.uk/UK-Market-Research-Reports/Facilities-
  Management.htm)

Sustainability Information

Building a dialogue with stakeholders on CSR reporting, GlobeScan Inc (www.globescan.
  com/csrr_analyzer.htm)
Building CEO Capital, Burson-Marsteller (www.ceogo.com/documents/Building_CEO_
  Capital_2003.pdf)
Business in the community helps members improve their impact on communities and the
  environment (www.bitc.org.uk)
CSR academy – a training and awareness site (www.csracademy.org.uk)
CSR business information (www.bsdglobal.com/issues/sr.asp)
CSR news (www.mallenbaker.net/csr/)
EC Green Paper Promoting a European framework for Corporate Social Responsibility
  (http://ec.europa.eu/employment_social/soc-dial/csr/greenpaper.htm)
EC Green Public Procurement (http://ec.europa.eu/environment/gpp/index.htm)
EIRIS ethical investment research service (www.eiris.org/)

Electronic reporting network for social, environmental, economic and corporate governance information (www.one-report.com)

EMAS (www.emas.org.uk/)

ENDS report online (www.endsreport.com)

Envirolink UK (www.envirolinkuk.org/)

Environment Council, The (www.the-environment-council.org.uk/)

Environmental news and information (www.edie.net/index.asp)

EU CSR Programme (www.eu.int/comm/enterprise/csr/index.htm)

Forum for the future (www.forumforthefuture.org.uk)

FTSE4good develops and maintains a series of global sustainable investment indices (www.ftse4good.com)

Gee Publishing (www.environment-now.co.uk)

Global Reporting Initiative is compiling sustainable reporting guidelines (www.globalreporting.org)

Green Futures magazine (www.greenfutures.org.uk)

Greenpeace (www.greenpeace.org)

Indoor Health and Productivity Project (www.ihpcentral.org)

Institute for social and ethical accountability (http://cei.sund.ac.uk/publications.htm)

Intergovernmental Panel on Climate Change (www.ipcc.ch)

The International Institute for Sustainable Development (IISD) (http://iisd.ca/)

James Lovelock's Gaia philosophy (www.ecolo.org/lovelock/whatis_Gaia.html)

Measuring Environmental Performance of Industry (MEPI) project (www.sussex.ac.uk/Units/spru/mepi/about/index.php)

The Natural Step (www.naturalstep.org.uk)

Next Step Consulting Ltd provides the Corporate Register – Environmental Reports (www.corporateregister.com)

Non-financial corporate reports (www.enviroreporting.com/)

Oxfam (www.oxfam.org)

Rainforest Action Network (www.ran.org)

SIGMA (www.projectsigma.com)

Social Accountability International developers of SA8000 (www.sa-intl.org/)

Socially responsible investment (www.socialfunds.com/)

SustainAbility (www.sustainability.com/)

Sustainability Alliance (www.sustainabilityalliance.org.uk)

SustainableBusiness.com (www.sustainablebusiness.com/)

Sustainable performance benchmarking (www.sustainability-performance.org/)

Tackling the Business of Sustainability (www.environmentawards.net/sage/)

UK Institute of Environmental Assessment and Management (www.iema.net/)

UNDP Human Development Report 2001 (hdr.undp.org/reports/global/2001/en/)

UN Food and Agriculture Organisation (www.fao.org/)

UN Global Compact (www.unglobalcompact.org)

UN Millennium Development Goals (www.un.org/milleniumgoals/)

Waste Indicators (http://themes.eea.eu.int/Environmental_issues/waste/indicators/)

World Business Council for Sustainable Development (WBCSD) (www.wbcsd.ch/)

World Economic Forum (www.weforum.org/)

World Resources Institute (www.globalforestwatch.org)

The Worldwatch Institute Report (www.worldwatch.org)

WWF (www.wwf.org.uk/)

Legislation

Australian Department of Environment and Heritage (www.deh.gov.au/)

European Environment Agency (www.eea.eu.int/)

European Union legislation and information (www.europa.eu.int/)

Hong Kong Environment Protection Department (www.epd.gov.hk/epd)

Montreal Protocol (www.unep.org/ozone/Montreal-Protocol/)

NetRegs is a web resource to help small companies understand environmental legislation (www.environment-agency.gov.uk/netregs)

The Environment Agency is interested in waste and pollution on construction sites, SUDS and operational water use (www.environment-agency.gov.uk)

US Environmental Protection Agency (www.epa.gov/epahome/)

General sustainable buildings

ASSER NL (www.asser.nl/EEL/)

Association of Certified Chartered Accountants' (ACCA) work on Corporate Social Responsibility and sustainability reporting (www.acca.org.uk)

Association of Environment Conscious Builders (AECB) (www.aecb.net)

Barbour Index (www.barbourexpert.com)

BetterBricks (www.betterbricks.com)

British Council for Offices (BCO) has produced advice on green roofs and fuel cells, and general guidance on 'Sustainability Starts in the Boardroom' and 'Sustainable Buildings are Better Business' (www.bco.org.uk)

British Property Federation (BPF) who have produced an Energy Guide for members (www.bpf.org.uk)

Building for Environmental and Economic Sustainability (BEES) National Institute of Standards & Technology (www.bfrl.nist.gov)

Building Green (www.buildinggreen.com)

Building Owners and Managers Association (www.boma.org)

Building Research Establishment (BRE) – see here for more information about BREEAM, EcoHomes, envest, Environmental Profiles, Green Guide, MaSC, post-occupancy evaluation, SMARTStart, SMARTWaste, whole life costing and much more (www.bre.co.uk)

Building Services Research and Information Association (BSRIA) (www.bsria.co.uk)

Business in the Environment (www.business-in-environment.org.uk)

Business Link (www.businesslink.org/)

Chartered Institute of Building (CIOB) (www.ciob.org.uk)

Chartered Institute of Building Services Engineers (CIBSE) (www.cibse.org)

Commission for Architecture and the Built Environment (CABE) (www.cabe.org.uk)

Considerate Constructors Scheme is a code of practice for improved construction sites (www.ccscheme.org.uk)

Constructing Excellence: bringing together Construction Best Practice (CBP) and Rethinking Construction (Movement for Innovation, The Housing Forum, Local Government Task Force) (www.constructingexcellent.org)

Construction Industry Council (CIC) is the representative forum for the industry's professional bodies, research organisations and specialist trade associations. The Happold

Lecture Series is available here, plus details about the Sustainable Development Committee (www.cic.org.uk)

Construction Industry Research and Information Association (CIRIA) – also has details of Construction Industry Environmental Forum (CIEF) and Construction Productivity Network (CPN) (www.ciria.org.uk)

Construction Industry Training Board ConstructionSkills (www.citb.org.uk)

Construction Resources (www.constructionresources.com)

Cool Roof Rating Council (www.coolroofs.org)

Co-operative Bank's Ethical Purchasing Index (www.greenconsumerguide.com/epi.php)

Croners (www.croner.net/index.html)

Design Quality Indicator (www.dqi.org.uk)

Environmental Building News (www.buildinggreen.com)

Environmental Building News (www.buildinggreen.com/index.html)

Environmental Design + Construction (www.edcmag.com)

Environmental Governance (www.environgov.co.uk)

Ethical Junction (www.ethical-junction.org)

EU Design Practices (www.unep.or.jp/ietc/sbc/index.asp)

FIT Buildings Network (www.theFBnet.com)

Global Alliance for Building Sustainability (GABS) Charter (http://webapps01.un.org/dsd/partnerships/public/partnerships/51.html)

Greener Buildings providing information on LEED (www.greenerbuildings.com/)

Greenguard Environmental Institute (www.greenguard.org)

Green Register of Construction Professionals (www.greenregister.org)

GreenSpec Building Green Inc. (www.buildinggreen.com)

House Builders Federation (www.hbf.co.uk)

HVCA (Heating Ventilation Contractors Association) (www.hvca.org.uk) Provides Standard Maintenance Specification for Mechanical Services in Buildings

Institute of Chemical Engineering/hamilton nashe (www.sustainability2000.org)

Institute of maintenance and building management (www.imbm.org.uk)

Institution of Civil Engineers (ICE) (www.ice.org.uk)

Integration of New and Renewable Energy in Buildings (INREB) (www.inreb.org)

Laboratories for the 21st Century (www.epa.gov/labs21century/)

Leadership in Energy and Environmental Design (www.usgbc.org/)

McDonough Braungart Design Chemistry (www.mbdc.com/)

Modular Design Construction (www.connet.org/uk/bp.jsp)

Movement for Innovation (M4I) (www.m4i.org.uk)

National House-Building Council (NHBC) (www.nhbc.co.uk)

New CRISP (nCrisp) (Construction Research and Innovation Strategy Panel) develops the research agenda for construction (www.crisp-uk.org.uk)

PlaceCheck a method for assessing the qualities of a place (www.udal.org.uk/placecheck.htm)

Planet GSA (http://hydra.gas.gov/planetgsa/)

Planning for Real is a community consultation method which is managed by the Neighbourhood Initiatives Foundation (www.nifonline.org.uk)

Prefabrication technologies (www.fabprefab.com)

Prince's Foundation is involved in a number of projects, and espouse 'Enquiry by Design' (www.princes-foundation.org)

Professional Association of Managing Agents (www.landcentre.ca/pama)

Property Environment Group (www.pegonline.net)

RICS Foundation (www.rics-foundation.org)

Rocky Mountain Institute (www.rmi.org)

Royal Institute of British Architects (RIBA) who have a Sustainable Futures group (www.architecture.com)

Royal Institution of Chartered Surveyors (RICS) (www.rics.org)

Royal Town Planning Institute (www.rtpi.org.uk)

SchoolWorks is a body working towards better school refurbishment and management (www.school-works.org)

Sponge – network for young professionals in sustainable construction (www.spongenet.org)

Strategic Forum – chaired by Peter Rogers. Web pages currently hosted by Construction Best Practice (www.cbpp.org.uk/acceleratingchange)

Support for Environmental Assessment and Management (SEAM) Project (www.environment awards.net/sage/)

Sustainable Buildings Industry Council (www.sbicouncil.org/)

Sustainable Building Sourcebook (www.greenbuilder.com/sourcebook)

Sustainable Data Bank (www.sd-eudb.net)

Sustainable Investment Research International, January 2002 (www.siricompany.com)

Sustainable materials sourcebook (www.greenbuilder.com/sourcebook/)

Water, Waste and Environment suppliers (www.water-waste-environment-marketplace. com/)

WellBuilt! a network for local authority professionals interested in more sustainable construction (www.wellbuilt.org.uk)

Whole Building Design Guide (www.wbdg.org/)

WWF has launched an initiative for One Million Sustainable Homes (www.wwf.org.uk/ sustainablehomes)

US government

Building Deconstruction Consortium focus on research, information dissemination and insti-tutionalization of deconstruction practices. To date, primary focus has been on military base deconstruction (www.buildingdeconstruction.org)

Federal Energy Management Program Federal Greening Toolkit (www.eere.energy. gov/femp/techassist/greening_toolkit/)

Federal Energy Management Program Greening Federal Facilities (www.eere.energy. gov/femp/techassist/green_fed_facilities.html)

Federal Energy Management Program Low-Energy Building Design Guidelines (www.eere. energy.gov/femp/prodtech/low-e_bldgs.html)

Federal Green Building (FedGB) Listserv covers over 300 Federal employees involved in green buildings (www.epa.gov/greenbuilding)

Federal Interagency Committee on Indoor Air Quality (CIAQ) (www.epa.gov/iaq/ciaq/)

Federal Network for Sustainability comprises mainly Western Regional offices of: Army, Navy, Air Force, DOE, EPA, NASA, NPS, USPS and Bonneville Power Administration (www.federalsustainability.org)

GSA's Design Excellence Program Guide (hydra.gsa.gov/pbs/pc/design_excell/)

Interagency Sustainability Working Group represents over a dozen Federal agencies, led by DOE's Federal Energy Management Program (FEMP) (www.eere.energy.gov/femp/ techassist/sustain_green.html)

Key rules and legislation affecting Federal facilities (www.eere.energy.gov/femp/resources/legislation.html)

Office of the Federal Environmental Executive (www.ofee.gov/)

Office of the Federal Environmental Executive waste minimisation programme (www.ofee.gov/wpr/wastestream.htm)

U.S. Department of Energy High Performance Buildings (www.eere.energy.gov/buildings/highperformance/)

U.S. Department of Housing and Urban Development's Guide to Deconstruction (www.huduser.org/publications/destech/decon.html)

U.S. Department of Interior's Guiding Principles of Sustainable Design (www.nps.gov/dsc/dsgncnstr/gpsd/)

U.S. Environmental Protection Agency (www.epa.gov/greenbuilding/green.htm)

U.S. Green Building Council (www.usgbc.org/)

Whole Building Design Guide (WBDG) is sponsored by DoD, DOE, EPA, FEMA, GSA, NASA, HHS/NIH, DVA, and other Federal agencies (www.wbdg.org)

UK government

Advisory Committee on Business and the Environment (www.defra.gov.uk/environment/acbe/index.htm)

DEFRA (www.construction.detr.gov.uk/cis/conmon/feb99/con6.htm, www.construction.detr.gov.uk/consult/eep/pdf/eep1.pdf,www.doingyourbit.org.uk,/www.environment.detr.gov.uk/epsim/indics/)

Housing Corporation (www.housingcorp.gov.uk) has funded Sustainability Works: a reference tool for sustainable housing (www.sustainabilityworks.org.uk) and Sustainable Homes which has produced many resources including Green Street (www.sustainablehomes.org.uk)

Sustainable Development Commission (www.sd-commission.gov.uk)

Sustainable Procurement Task Force (www.sustainable-development.gov.uk/government/task-forces/procurement/index.htm)

UK Government Procurement (www.ogcbuyingsolutions.gov.uk)

UK Greening Government (www.environment.detr.gov.uk/greening/gghome.htm)

Index

Note: page numbers in **bold** refer to tables while those in *italics* are for figures.